/

PROBLEMS in CHEMICAL THERMODYNAMICS WITH SOLUTIONS

PROBLEMS in CHEMICAL THERMODYNAMICS with SOLUTIONS

Maka Aleksishvili | Shota Sidamonidze

Tbilisi State University, Georgia

Translated from Georgian into English by
Professor Marina Rukhadze

World Scientific
New Jersey • London • Singapore • Hong Kong

Published by

World Scientific Publishing Co. Pte. Ltd.

P O Box 128, Farrer Road, Singapore 912805

USA office: Suite 1B, 1060 Main Street, River Edge, NJ 07661

UK office: 57 Shelton Street, Covent Garden, London WC2H 9HE

British Library Cataloguing-in-Publication Data
A catalogue record for this book is available from the British Library.

PROBLEMS IN CHEMICAL THERMODYNAMICS, WITH SOLUTIONS

ISBN 981-238-076-0

This book is printed on acid-free paper.

Printed in Singapore by Mainland Press

Contents

Preface

The comprehension and assimilation of some main questions and concepts of physical chemistry create certain difficulties for students. The nature of entropy and thermodynamic potentials, the meaning of thermal effect, the analogy and difference between heat exchange and work belong to the problems of this kind.

Our long-term practice indicates the complexity of perception of these questions by traditional methods. Superficial knowledge causes the loss of interest regarding to the subject. The goal of proposed work was the creation of the clear view in students about the basic topics of chemical thermodynamics. For achievement of this purpose we did not avoid the use of nontraditional methods of exposition. This results in the original structure of this book. Each chapter includes more or less extended theoretical part and problems with solutions. The essential minimum of the actual knowledge is given in the theoretical part. This minimum is explained by using the methods and views, which facilitate the thorough understanding of the topic e.g. entropy is interpreted by using of statistical thermodynamics, which is more clear then classical approach. The change of entropy of surroundings and "universe", the "noncompensated" heat of Clausius, the entropy production, the "lost" work are considered with the same purpose. The existence of internal connection between thermodynamic potentials and production of entropy is shown. The method of generalized forces is utilized e.g. in order to establish the analogy between heat exchange and work.

To overcome the loading of the material, some questions are not introduced in the theoretical part, but are discussed in the problems e.g. the difference between heat exchange and work, relation between heat exchange and entropy, which are explained by means of variations of the population and heights of energy levels of the system. In problems of the "Great fluctuations" is shown, that the second law of thermodynamics possesses not absolute but statistical character. There is shown also that the probability of the spontaneous proceeding of the unnatural process is practically equal to zero; period, which is required for its realization is incomparably large with the existence time of the universe. The ideality and unideality of a gas is associated with classical and quantum modes of the process proceeding.

Some problems are discussed in view of both cognitive and historical intention e.g. discussion between Boltzman and his opponents, Gibbs's paradox etc.

The relation between the integrating constant of entropy and expressions of statistical thermodynamics is deduced in problems devoted to Gibbs' paradox. It is shown that Gibbs's paradox is stipulated by neglecting of concentration in the classical expression of entropy. The expressions of chemical constants and their numerical values are determined from Sacure-Tetrode' formulae, which were deduced early by using of statistical and quantum views.

We tried to establish the connections between different chapters e.g. the expansion processes are considered by using both the first and second laws of thermodynamics. The alteration of thermodynamic potential is discussed in connection with changes in entropy of system, surroundings and "universe".

The questions of practical importance are also considered in the proposed book, viz. the principles of working and peculiarities of automobile, refrigerator, air conditioner and heat pump (reversed air conditioner).

Problems are solved scrupulously. They are accompanied by remarks, commentaries and explanations of different character. The conclusions are given at the close of the tables (which combines the results of several problems). This promotes to comprehense the basic meaning of problems and to interpret the material on the whole.

The SI units are mainly used in the proposed book. However, sometimes the CGSM and nonsystemic units are utilized e.g. calorie, atmosphere, liter.

We think because of the abovementioned novelties the book will not be insured from the deficiencies. The authors will receive the critical remarks with gratitude.

M. Aleksishvili
Sh. Sidamonidze

1. The first Law of Thermodynamics

Theoretical Part

The first law of thermodynamics represents a private aspect of the fundamental law of nature — the law of energy constancy for thermodynamic systems i.e. for systems, where the heat exchange and work take place. Its analytical expression is:

$$dU = \delta q + \delta A \qquad (1.1)$$

where dU is the infinitesimal change of internal energy, δq represents the infinitisimal amount of heat, δA—the infinitesimal amount of work.

The work may be presented as sum of the two items:

$$\delta A = \delta A_{\text{exp.}} + \delta A' \qquad (1.2)$$

where $\delta A_{exp.}$ is work of expansion and $\delta A'$ represents so–called useful work (electrical, chemical, etc).

When only external pressure acts on the system, $\delta A' = 0$ and

$$\delta A_{\text{exp}} = -P_{ex}dV \qquad (1.3)$$

where P_{ex} is external pressure and dV represents the infinitesimal change of volune of the system.

In this case the first law of thermodynamics assumes the following form:

$$dU = \delta q - P_{ex}dV \qquad (1.4)$$

For finite change:

$$\Delta U = q + A_{exp.} \tag{1.5}$$

$$A_{exp.} = -\int_{V_1}^{V_2} P_{ex} dV \tag{1.6}$$

$$\Delta U = q - \int_{V_1}^{V_2} P_{ex} dV \tag{1.7}$$

When external pressure is constant, then it is followed from expression (1.6):

$$A_{exp.} = -P_{ex} \Delta V \tag{1.8}$$

In case of $P_{ex} = const$ (expansion in a vacuum), $A_{exp.} = 0$

Distinction between U, q and A

Internal energy represents the sum of energies of all kind for all particles. It includes translational, rotational, vibrational energies, energy of motion of electrons in atom, nuclear energy, etc. The kinetic and potential energies of system as a whole, are not considered in this sum. Calculation of absolute value of internal energy is impossible. Determination of its variation is possible only : $\Delta U = U_2 - U_1$. If internal energy of system increases as a result of process, $\Delta U > 0$ and vice versa. Internal energy is a state function. It determines the state of system unambiguously. Therefore change of internal energy depends on the initial and final states of system only and is independent on the way of process. The finite change of internal energy is expressed so:

$$\Delta U = \int_{U_1}^{U_2} dU$$

Heat and work do not represent any form of energy. They are qualitative and quantitative characteristics of energy transfer by one or another mode. q and A do not characterize the state of system: the system possesses internal energy in one or another state, but has no work or heat. A and q reveal in the process only, when energy exchange between system and surroundings is performed by their owing. Hence, A and q do not represent a state function: they characterize not a state, but a process. Their values are depended on the way of process. The finite amount of A and q are expressed so:

$$A = \int_{way} \delta A \qquad \text{and} \qquad q = \int_{way} \delta q$$

Signs of q and A

The signs of heat and work are selected according to increase or decrease of internal energy of system in the given process.

When system releases heat or performs work, its internal energy decreases. Therefore the negative signs are given to heat and work:

$q < 0$ (system releases heat)

$A < 0$ (system fulfils work)

If system receives heat from surroundings or surroundings fulfils work on system, the internal energy of system increases. Heat and work have positive values:

$q > 0$ (system gains heat)

$A > 0$ (work is fulfilled on system).

Enthalpy

The use of enthalpy — the function derived from internal energy is convenient in many cases (especially in conditions of constant pressure):

$$H = U + PV \qquad (1.9)$$

From expression (1.9) is obtained:

$$dH = dU + PdV + VdP + dPdV \tag{1.10}$$

For finite change may be written:

$$\Delta H = \Delta U + P\Delta V + V\Delta P + \Delta P\Delta V = \Delta U + P_2V_2 - P_1V_1 \tag{1.11}$$

Because $\Delta P\Delta V$ has a minute value, it is often neglected. Then

$$\Delta H = \Delta U + P\Delta V + V\Delta P \tag{1.12}$$

Enthalpy is a state function similar to internal energy.

Heat capacity
If temperature of system due to heat gain increases from T_1 to T_2, then quantity

$$c_{av.} = \frac{q}{T_2 - T_1}$$

is called average heat capacity of system.

Heat capacity corresponded to the infinitesimal change of temperature is termed real heat capacity:

$$c = \frac{\delta q}{dT}$$

The value of both heat capacity and heat, are depend on the way of process. Heat capacity in conditions of constant volume (c_V) and constant pressure (c_p) is considered mainly:

$$c_V = \frac{\delta q_v}{dT} = \frac{dU}{dT} \tag{1.13}$$

$$c_p = \frac{\delta q_p}{dT} = \frac{dH}{dT} \tag{1.14}$$

From these equations:

$$dU = \delta q_V = nc_V dT \tag{1.15}$$

and

$$dH = \delta q_p = nc_p dT \tag{1.16}$$

c_V and c_P are related with each other in the following way:

$$c_P - c_V = nR \tag{1.17}$$

where n is number of moles and R represents gas constant.

Equilibrium and nonequilibrium (reversible and irreversible) processes

Thermodynamic quantities may be considered from various standpoint. Let us group them as so–called generalized forces and generalized coordinates. The generalized forces (common designation F_k) represent: force F, pressure P, chemical potential μ, electric potential φ, etc. The generalized coordinates (common designation x_k) are: length l, volume V, number of moles n, charge e, etc. The generalized coordinates are varied under the influence of corresponded to them generalized forces; the generalized work is fulfilled as a result of this:

$$\delta A_k = F_k dx_k$$

Namely,

$$\delta A_{mech.} = Fdl$$
$$\delta A_{exp.} = PdV$$
$$\delta A_{chem.} = \sum \mu_i dn_i$$
$$\delta A_{elec.} = \varphi de$$

The elementary act is termed reversible, when its inducing reason is infinitesimal difference between generalized forces, by which system effects the surroundings and vice versa, the surroundings acts on the system:

$$F_{in} = F_{ex} + dF$$

(subscript "in" corresponds to system, but "ex" to surroundings).

The infinitesimal change of generalized force (dF) may result in the change of sign of generalized coordinate's increment (dx). Let us consider the gas expansion for example.

Let imagine, that gas is placed in cylinder with piston. The infinitesimal weights are on the piston, by means of which surroundings performs pressure on the system (P_{ex}). At the equilibrium $P_{ex} = P_{in}$. If we take away one weight, internal pressure of system becomes higher by infinitesimal quantity than external pressure ($P_{in} > P_{ex}$) and elementary (infinitesimal) act of expansion will occur. Hence, internal and external pressures become equal again and equilibrium is established. If now on the contrary, we increase external pressure by infinitesimal quantity, then $P_{ex} > P_{in}$ and elementary act of compression takes place. As a result P_{in} increases, becomes equal to P_{ex} and the equilibrium is established.

The finite process, constituted with infinite number of monodirectional equilibrium elementary acts is called equilibrium (reversible) process. Such process proceeds infinite slowly and its realization requires infinite time. Equality of generalized forces on the test surface (more precisily, the existence of infinitesimal difference between them) is the requirement of equilibrium proceeding of process.

The expansion is reversible (equilibrium), when external pressure varies during process so, that the condition $P_{in} = P_{ex} + dP$ or more simply $P_{in} \approx P_{ex}$ is fulfilled in all stages. Taking into account this condition, from expressions (1.3) and (1.6) is obtained:

$$\delta A_{revers.} = -P_{in}dV \qquad (1.18)$$

and

$$A_{revers.} = -\int_{V_1}^{V_2} P_{in}dV \qquad (1.19)$$

Expressions (1.18) and (1.19) may be used in reversible process only. Substitution of P_{ex} by P_{in} simplifies the determination of A, because it gives the possibility to use the ideal gas equation.

The expansion is nonequilibrium (irreversible) if the finite difference between internal and external pressures exists: $P_{in} = P_{ex} + \Delta P$. In this case the process proceeds nonequilibriumly, with finite rate and in the finite time. The expansion at constant external pressure ($P_{ex} = const$) is often considered in thermodynamics. Such expansion is irreversible, if internal pressure changes during the process, but external pressure is constant and the condition of reversibility: $P_{in} \approx P_{ex}$ is not realized.

The generalized expressions (1.3) and (1.6) will be used for irreversible processes; they will transform into one or another form in concrete conditions (expansion at constant external pressure, expansion in a vacuum, isochoric, isobaric, isothermal, adiabatic expansions).

The work of reversible expansion exceeds the irreversible one. Indeed,

$$A_{revers.} = -\int_{V_1}^{V_2} P_{in} dV \tag{1.19}$$

$$A_{irrev.} = -\int_{V_1}^{V_2} P_{ex} dV \tag{1.6}$$

At the same time the condition of irreversible expansion is: $P_{in} > P_{ex}$. It follows, that

$$\left| A_{revers.} \right| > \left| A_{irrev.} \right|$$

The temperature performs the function of generalized force and entropy S – the function of generalized coordinate during heat exchange:

$$\delta q = TdS \tag{1.20}$$

The questions devoted to heat exchange are considered in chapter 3. Now the identical form of expressions (1.3) and (1.20) should be mentioned:

work and heat analogously are related to the intensive (P, T) and extensive (V, S) parameters or to the generalized forces and generalized coordinates. This similarity is not random and creates a certain basis of heat—work equivalence principle.

The use of first law of thermodynamics in various processes

Let us consider expansion of gas in various processes (in all cases the number of moles is constanr: $n = const$).

1) Isochoric process $(V = const)$

$$A = -\int_{V_1}^{V_2} P_{ex} dV = 0 \qquad (1.21)$$

$$\Delta U = q_v = n\bar{c}_v (T_2 - T_1) \qquad (1.15)$$

$$\Delta H = \Delta U + \Delta(PV) \qquad (1.22)$$

From the ideal gas equation:

$$\Delta(PV) = \Delta(nRT) \qquad (1.23)$$

When $V, n = const$:

$$\Delta(PV) = V\Delta P$$

and

$$\Delta(nRT) = nR\Delta T$$

$$V\Delta P = nR\Delta T$$

Thus

$$\Delta H = \Delta U + V\Delta P \qquad (1.24)$$

and

$$\Delta H = \Delta U + nR\Delta T \qquad (1.25)$$

2) Isobaric process (*P=const*)

During the whole process of isobaric expansion $P_{in} = P_{ex} = const$. Increasing of temperature represents a motive force of process. Expansion interrupts when system occupies that volume, which corresponds to given values of P, T and n.

Isobaric expansion must be distinguished from so–called "expansion opposite constant external pressure". In this last case $P_{in} > P_{ex}$ and equalization of internal and external pressures takes place at the end of process only. But at isobaric expansion $P_{in} = P_{ex}$ during the whole process.

Work of isobaric expansion is expressed so:

$$A = -\int_{V_1}^{V_2} P_{ex} dV \qquad (1.6)$$

While $P_{in} = P_{ex} = const$ at isobaric expansion, therefore

$$A = -P_{ex}\Delta V = -P_{in}\Delta V$$

or simplier

$$A = -P\Delta V \qquad (1.26)$$

In the conditions $P, n = const$ from the ideal gas equation follows:

$$P\Delta V = nR\Delta T \qquad (1.27)$$

and

$$A = -nR\Delta T \qquad (1.28)$$

As is known,

$$\Delta U = n\bar{c}_V \Delta T \qquad (1.15)$$

$$\Delta H = \Delta U + P\Delta V + V\Delta P \qquad (1.12)$$

At $P = const$

$$\Delta H = \Delta U + P\Delta V \qquad (1.29)$$

Taking into consideration expression (1.27), it is obtained:

$$\Delta H = \Delta U + nR\Delta T \qquad (1.30)$$

If take into account, that $\Delta U = q_v$ and $\Delta H = q_p$, then from (1.29) and (1.30) we obtain:

$$q_P = q_V + P\Delta V \qquad (1.31)$$

$$q_P = q_V + nR\Delta T \qquad (1.32)$$

3) Isothermal process $(T = const)$
Internal energy of ideal gas is a function of temperature only. Hence in conditions of $T = const$,

$$\Delta U = 0 \qquad (1.33)$$

$$\Delta H = \Delta U + \Delta(PV) \qquad (1.22)$$

In the conditions $T, n = const$

$$\Delta(PV) = \Delta(nRT) = 0$$

Thus,
$$\Delta H = 0 \qquad (1.34)$$

It follows from expressions (1.5) and (1.33), that in isothermal process

$$A = -q \qquad (1.35)$$

At reversible isothermal expansion:

$$A_{revers.} = -\int_{V_1}^{V_2} P_{ex}dV = -\int_{V_1}^{V_2} P_{in}dV = -\int_{V_1}^{V_2}\frac{nRT}{V}dV =$$

$$= -nRT \int_{V_1}^{V_2} \frac{dV}{V} = -nRT \ln \frac{V_2}{V_1} = nRT \ln \frac{P_2}{P_1} \qquad (1.36)$$

From the comparison of expressions (1.35) and (1.36) it is obtained:

$$q_{revers.} = nRT \ln \frac{V_2}{V_1} = nRT \ln \frac{P_2}{P_1} \qquad (1.37)$$

4) Adiabatic process

Heat exchange between system and surroundings is excluded in adiabatic process:

$$q = 0 \qquad (1.38)$$

Therefore according to first principle of thermodynamics

$$\Delta U = A \qquad (1.39)$$

It follows from expressions (1.15) and (1.39), that in adiabatic process

$$\Delta U = A = n\bar{c}_v (T_2 - T_1) \qquad (1.40)$$

or

$$A = \frac{nR}{\chi - 1} (T_2 - T_1) \qquad (1.41)$$

where

$$\chi = \frac{\bar{c}_p}{\bar{c}_v} \qquad (1.42)$$

and

$$\chi - 1 = \frac{R}{\bar{c}_v} \qquad (1.43)$$

System fulfils work at adiabatic expansion owing to its internal energy. Hence internal energy of system decreases, which causes reducing of temperature also: $\Delta T = T_2 - T_1 < 0$.

Surroundings fulfils work on system at adiabatic compression i.e. gives it energy. Hence internal energy of system increases; $\Delta U > 0$ and $\Delta T = T_2 - T_1 > 0$.

The change of enthalpy in adiabatic process is:

$$\Delta H = \Delta U + \Delta(PV) = \Delta U + \Delta(nRT) = \Delta U + nR\Delta T =$$
$$= n\bar{c}_v\Delta T + nR\Delta T = n\Delta T(\bar{c}_v + R) = n\bar{c}_p\Delta T \qquad (1.44)$$

Expressions (1.38÷1.44) are correct for both reversible and irreversible adiabatic expansions.

a) Reversible adiabatic expansion

The essential condition for reversible proceeding of process is $P_{ex} = P_{in}$. Taking into account this circumstance and by using of ideal gas expression, the equations of adiabate are derived:

$$PV^{\chi} = const \qquad (1.45)$$

$$TV^{\chi-1} = const \qquad (1.46)$$

$$T^{\chi}P^{1-\chi} = const \qquad (1.47)$$

Besides equations (1.40) and (1.41) the following expressions may be used for determination of the work of reversible adiabatic expansion

$$A = \frac{P_2V_2 - P_1V_1}{\chi - 1} \qquad (1.48)$$

$$A = \frac{P_1V_1}{\chi - 1}\left[\left(\frac{P_2}{P_1}\right)^{\frac{\chi-1}{\chi}} - 1\right] \qquad (1.49)$$

$$A = \frac{nRT_1}{\chi - 1}\left[\left(\frac{P_2}{P_1}\right)^{\frac{\chi-1}{\chi}} - 1\right] \qquad (1.50)$$

$$A = \frac{P_1 V_1}{\chi - 1}\left[\left(\frac{V_1}{V_2}\right)^{\chi-1} - 1\right] \tag{1.51}$$

b) Irreversible adiabatic expansion
($P_{ex} = const$)
At expansion opposite to constant external pressure

$$A_{irrev.} = -P_{ex}\Delta V \tag{1.8}$$

At the same time, at adiabatic expansion of any kind

$$A = n\bar{c_V}\Delta T \tag{1.40}$$

Let us equalize expressions (1.8) and (1.40):

$$-P_{ex}\Delta V = n\bar{c_V}\Delta T$$

From this expression:

$$\frac{\Delta T}{\Delta V} = -\frac{P_{ex}}{n\bar{c_V}} \tag{1.52}$$

Expression (1.52) is used to determine parameters of irreversible adiabatic process.

Remark: Expression (1.40) is correct for both reversible and ireversible adiabatic process. But at the same initial conditions $\Delta T_{rev.} \ne \Delta T_{irrev.}$ and therefore $A_{rev.} \ne A_{irrev.}$.

Problems

Problem № 1.1

3 mole of Ar was heated by 5^0 in closed reactor with volume 1 m^3. What is the work fulfilled in this process?

Solution:

Work of expansion is equal to:

$$A_{exp.} = -\int_{V_1}^{V_2} P_{ex} dV \qquad (1.6)$$

Because reactor is closed, $dV = 0$ and hence $A = 0$. Thus, work of expansion is not fulfilled in closed reactor.

Problem № 1.2.

1 mole of crypton is placed in the cylinder with a piston at a constant internal and external pressure (5 atm); 416 J energy expends on its heating.

1) What are the values of ΔV and ΔT in the process of crypton expansion?

2) What is the work fulfilled in this process?

($\bar{c}_p$ of crypton equals to 20.78 J / K×mol).

Solution:

1) At heating under constant pressure:

$$q_p = n\bar{c}_p \Delta T \qquad (1.16)$$

Hence

$$\Delta T = \frac{416}{1 \times 20.78} = 20.02^0 \text{ K}$$

According to ideal gas equation under $P, n = const$ conditions:

$$P\Delta V = nR\Delta T \qquad (1.27)$$

and

$$\Delta V = \frac{nR\Delta T}{P} = \frac{1 \times 0.082 \times 20}{5} = 0.328 \ L$$

2) Work, fulfilled at constant pressure is equal:

$$A = -P\Delta V = -5 \times 0.33 = -1.64 \ L \times atm = -166.2 \ J.$$

Problem № 1.3

1 mole of crypton presents in cylinder, corked up tightly under 5 atm pressure. It undergoes the heating in the same temperature range as in problem № 1.2.

What amount of heat receives the system?

What is the work, fulfilled by the system?

Solution:

The process proceeds in conditions of constant volume ($V = const$), because cylinder is closed compact . (At the same time, $P \neq const$, while pressure of gas increases at heating.).

According to expression (1.15):

$$q_v = n\bar{c}_v \ \Delta T = 1 \times 12.47 \times 20 = 249.4 \ J$$

Work of expansion is not fulfilled in conditions of constant volume:

$$A = -\int_{V_1}^{V_2} P_{ex} dV = 0 \qquad (1.21)$$

Remark: As is seen from comparison of problems № 1.2 and 1.3,

$q_p > q_v$ by heating of system with the same initial state   in the same temperature range, but in different conditions ($P = const$ and $V = const$). This is caused by following: heat at constant pressure expends on both heating and work of expansion, but heat at constant volume expends on heating only. Therefore, $q_p = q_v + P\Delta V$ or $q_p = q_v + nR\Delta T$.

Problem № 1.4

85.77 kJ heat is absorbed at the evaporation of 100 g ethanol under 1 atm pressure and 6.15 kJ work is fulfilled.

1) What is the molar heat of evaporation of ethanol?
2) What amount of work is fulfilled at the evaporation of 1 mole ethanol?
3) What is the molar volume of vapor of ethanol?

Solution:

$$n_{ethanol} = \frac{100}{46} = 2.17 \ moles$$

1) The evaporation proceeds under constant pressure ($P = 1 \ atm$). Therefore,

$$q_p = n\Delta H = n\,\overline{\Delta H}_{evapor.}$$

From here on

$$\overline{\Delta H}_{evapor.} = \frac{q_p}{n} = \frac{85770}{2.17} = 39525 \ J = 39.525 \ kJ.$$

2) 6.15 *kJ* work is fulfilled at the evaporation of 100 g ethanol according to condition of problem. Then work fulfilled at evaporation of 1 mole (46 *g*) of ethanol is equal:

$$A = -\frac{46}{100} \times 6.15 = -2.829 \ kJ$$

3) As was calculated above, work fulfilled at evaporation of 1 mole ethanol is:

$$A = -P\Delta \overline{V} = -P(\overline{V}_{vap.} - \overline{V}_{liq.}) \approx -P\overline{V}_{vap.} = -2.835 \ kJ.$$

Hence, the molar volume of vapor is:

$$\overline{V}_{vap.} = \frac{2835 \times 9.867 \times 10^{-3}}{1} = 27.97 \ L$$

Problem № 1.5

The reaction:

$$2 \ KNO_3 \rightarrow 2 \ KNO_2 + O_2$$

proceeds in a cylinder with piston under 1 atm pressure and at 300^0C temperature. Piston with cross–section 200 cm^2 shifts at certain distance and 405.4 J work is fulfilled as a result of reaction.

1) What is the distance of shift of a piston?
2) How many moles of KNO_3 decompose as a result of reaction?

Solution:

1) Expansion proceeds irreversibly opposite the constant external pressure. Work, fulfilled by this process is expressed so:

$$A_{exp.} = -P_{ex}\Delta V \qquad (1.8)$$

From here on

$$\Delta V = -\frac{A}{P_{ex}} = \frac{405.4}{1 \times 101.32 \times 10^3} = 4 \times 10^{-3} \ m^3 = 4 \ L = 4000 \ cm^3$$

On the other hand, a change of volume at the shift of piston is:

$$\Delta V = s \times l$$

where s is cross–section of piston and l represents a distance at which a piston shifts.

$$l = \frac{\Delta V}{s} \qquad\qquad (a)$$

Let us enter the values of ΔV and s into expression (a):

$$l = \frac{4000}{200} = 20 \; cm$$

2) If the initial volume of system is not taken into account, then

$$\Delta V = V_2 - V_1 \approx V_2 = V_{O_2} = 4 \; L.$$

In order to establish the number of moles of oxygen, we refer to ideal gas equation:

$$PV = nRT$$

$$n = \frac{PV}{RT} = \frac{1 \times 4}{0.082 \times 573} = 0.085 \; mol$$

Equality of reaction is:

$$2 \; KNO_3 \rightarrow 2 \; KNO_2 + O_2$$

$$1 \; \text{mole } O_2 \;\text{————}\; 2 \; \text{mole } KNO_3$$

$$0.085 \; \text{mole } O_2 \;\text{————}\; x \; \text{mole } KNO_3$$

$$x = \frac{0.085 \times 2}{1} = 0.17 \; mol \; KNO_3$$

Problem № 1.6

A definite force is affected on piston instead of 1 atm pressure in system, described in previous problem.

What is the value of this force (kg), if the same amount of work is fulfilled in system as in previous case?

Solution:

$$A = Fl \qquad\qquad (a)$$

where F is the force, which affects on piston and l represents distance, corresponding to shift of piston.

Hence,

$$F = \frac{A}{l} \qquad\qquad (b)$$

Work 405.4 J is fulfilled at shift of piston with cross–section 200 cm^2 to 20 cm (0.2 m) upwards, as it is given in condition of problem № 1.7. Since 1 $kg \times m$ corresponds to 9.8 J,

$$A = \frac{405.4}{9.8} = 41.36 \ kg \times m$$

and

$$F = \frac{A}{l} = \frac{41.36}{0.2} = 206.8 \ kg$$

Problem № 1.7

Two ideal gases expand isothermal and reversible from V_1 to V_2. The mole number and temperature of gases are identical. Heat capacity of one gas twice exceeds heat capacity of the other gas.

What is the work fulfiled by each gas at the expansion?

Solution:

Work of isothermal reversible expansion is expressed so:

$$A = -nRT\ln\frac{V_2}{V_1} \qquad (1.36)$$

As is shown from expression (1.36), work does not depend on heat capacity. It is the same for both gases because, according to condition of problem, n, T, V_1 and V_2 have equal values in both gases.

Problem № 1.8

The ideal gas is compressed isothermally and reversibly at 400 K from 1 m^3 to 0.5 m^3. 9200 J heat is evolved at compression.

What is the work fulfilled and how many moles of gas compresses in this process?

Solution:

The internal energy does not change at isothermal compression of ideal gas:

$$\Delta U = 0.$$

Therefore

$$q = -A \qquad (1.35)$$

On the other hand,

$$q = -A = nRT\ln\frac{V_2}{V_1} \qquad (1.37)$$

Hence

$$n = \frac{q}{RT\ln\frac{V_2}{V_1}} = \frac{-9200}{8.31 \times 400\ln\frac{0.5}{1}} = \frac{-9200}{-2304} = 4\,mole$$

Problem № 1.9

2 moles of argon present in the cylinder with piston at room temperature and under 2 atm pressure. The system expands isothermally: 1) reversible to 1 atm pressure; 2) irreversible opposite to 1 atm pressure.

What is the work, fulfilled at these processes?

Solution:

Two moles of argon at ambient temperature and under 2 atm pressure occupies a volume:

$$V_1 = \frac{nRT}{P_1} = \frac{2 \times 0.082 \times 298}{2} = 24.44 \ L$$

Volume of argon after expansion is:

$$V_2 = \frac{298}{273} \times 2 \times 22.4 = 48.903 \ L$$

1) According to expression (1.36), work of reversible isothermal expansion is:

$$A_{revers.} = nRT \ln \frac{P_2}{P_1} = 2 \times 8.31 \times 298 \ln \frac{1}{2} = -3433 \ J$$

2) The work, fulfilled at irreversible isothermal expansion is:

$$A_{irrev.} = -P_{ex}\Delta V \qquad\qquad (1.8)$$

$$\Delta V = V_2 - V_1 = 48.90 - 24.45 = 24.45 \ L$$

$A = -1 \times 24.45 = -24.45 \ L \times atm = -24.45 \times 101.34 = -2476.7 \ J.$

Remark : As is shown from condition of problem, the values of fulfilled work are drastically different : $A_{revers.} = -3433 \ J$, $A_{irrevers.} = -2476 J.$

The reason of this difference is following: work of any process is equal to:

$$A_{exp.} = -\int_{V_1}^{V_2} P_{ex} dV \qquad (1.6)$$

At reversible expansion

$$P_{in} = P_{ex} \qquad (a)$$

and

$$A_{revers.} = -\int_{V_1}^{V_2} P_{in} dV \qquad (1.19)$$

At irreversible expansion

$$A_{irrevers.} = -\int_{V_1}^{V_2} P_{ex} dV \qquad (b)$$

During the whole process of isothermal expansion

$$P_{in} > P_{ex} \qquad (c)$$

From the comparison of expressions: (1.19), (b) and (c) is shown, that work fulfilled by system in reversible process exceeds work of irreversible process (in case of the same initial and final states).

Problem № 1.10

64 g of O_2 is present at 25^0C and 1 atm pressure. Molar heat capacity of oxygen is equal to 29.33 J / K×mol. Find the values of q, A and ΔU:

1) At doubling of gas volume in: a) reversible isothermal, b)irreversible isothermal, c)isobaric processes;

2) At isochoric doubling of gas pressure.

Tabulate the obtained results.

Solution:

$$n_{O_2} = \frac{64}{32} = 2 \; mole$$

1a) Work of reversible isothermal expansion is expressed so:

$$A = - nRT\ln\frac{V_2}{V_1} \qquad (1.36)$$

According to condition of problem, $\frac{V_2}{V_1} = 2$. Then

$$A = -nRT\ln2 = -2 \times 8.31 \times 298 \times 0.693 = -3432 \; J$$

In isothermal process $\Delta U = 0$ and

$$q = -A = 3432 \; J$$

1b) At irreversible isothermal expansion

$$A = -P_{ex}\Delta V \qquad (1.8)$$

Expansion ends, when internal pressure of system and external pressure are flattened: $P_2 = P_{ex}$.
In isothermal process:

$$\frac{P_2}{P_1} = \frac{V_1}{V_2}$$

According to condition of problem

$$\frac{V_1}{V_2} = \frac{1}{2}$$

Hence

$$\frac{P_2}{P_1} = \frac{1}{2}$$

and

$$P_2 = \frac{P_1}{2} = \frac{1}{2} \; atm$$

Thus, $P_2 = P_{ex} = 0.5 \; atm.$

$$\Delta V = V_2 - V_1 = 2V_1 - V_1 = V_1$$

From the ideal gas equation:

$$V_1 = \frac{nRT}{P_1} = \frac{2 \times 0.082 \times 298}{1} = 48.872 \; L$$

The fulfilled work is equal to:

$$A = -P_{ex}\Delta V = -0.5 \times 48.872 = -24.436 \; L \times atm =$$

$$= -24.436 \times 101.34 \; J = -2476 \; J.$$

Because the process is isothermal, $\Delta U = 0$ and $q = -A = 2476 \; J.$

1c) At isobaric expansion

$$q_p = n\bar{c}_p(T_2 - T_1) \qquad (1.16)$$

In conditions of constant pressure,

$$\frac{T_2}{T_1} = \frac{V_2}{V_1}$$

According to condition of problem, $\dfrac{V_2}{V_1} = 2$

Thus, $T_2 = 2T_1 = 2 \times 298 = 596$ K

Hence

$$q_p = 2 \times 29.33 \, (2T_1 - T_1) = 2 \times 29.33 \times 298 = 17481 \, J.$$

Fulfilled work is:

$$A = - P\Delta V \qquad (1.26)$$

From the ideal gas equation in conditions of $P, n = const$:

$$P\Delta V = nR\Delta T \qquad (1.27)$$

Thus,

$$A = - nR\Delta T = - 2 \times 8.31 \times 298 = -4952 \, J.$$

According to the first law of thermodynamics,

$$\Delta U = q + A = 17481 - 4952 = 12529 \, J.$$

2) Work is not fulfilled at isochoric doubling of gas pressure:

$$A = - \int P_{ex} dV = 0 \qquad (1.21)$$

Therefore the change of internal energy is:

$$\Delta U = q_V = n \bar{c}_V \, (T_2 - T_1) \qquad (1.15)$$

According to expression (1.17):

$$\bar{c}_V = \bar{c}_P - R = 29.33 - 8.31 = 21.02 \, J / K \times mol$$

In conditions $V, n = const$:

$$\frac{T_2}{T_1} = \frac{P_2}{P_1}$$

According to condition of problem: $P_2 = 2P_1$. Thus,

$$T_2 = 2T_1 = 596 \text{ K}$$

and

$$T_2 - T_1 = 2T_1 - T_1 = T_1 = 298 \text{ K}$$

Let us introduce the values of $\bar{c}_V$ and $(T_2 - T_1)$ in expression (1.15):

$$\Delta U = q_V = 2 \times 21.02 \times 298 = 12529 \text{ J}.$$

Let tabulate the obtained results (Table 1.1)

Table 1.1. Characteristic values for expansions of various kind.

(2 mole O_2, $T_1 = 298$ K, $V_1 = 48.9$ L, $P_1 = 1$ atm)

Kind of process	T_2, K	V_2, L	P_2, atm	A, kJ	q, kJ	ΔU, kJ
Isobaric expansion	596	97.7	1	− 4.952	17.481	12.529
Isochoric increasing of pressure	596	48.9	2	0	12.529	12.529
Isothermal expansion (reversible)	298	97.7	0.5	− 3.432	3.432	0
Isothermal expansion (irreversible)	298	97.7	0.5	− 2.476	2.476	0

As the table indicates, the system gains more heat from surroundings and fulfills more work at reversible isothermal expansion, than in irreversible one. Its reason was already considered in problem № 1.9.

The comparison of values of isobaric and isochoric processes is interesting. Despite difference in the final state of system in these processes (viz. the values of V_2 and P_2), ΔU has the same value. This is stipulated by the fact, that T_1 and T_2 are identical in both isobaric and isochoric processes; and internal energy of ideal gas is temperature function only.

Problem № 1.11

Temperature decreases by 100^0 at adiabatic expansion of 2 mole monoatomic gas. What is the work fulfilled in this process?

Solution:

At adiabatic expansion $q = 0$. Therefore

$$\Delta U = A = n\bar{c_v}(T_2 - T_1) \qquad (1.40)$$

$\bar{c_v} = 12.47 \; J/K \times mol$ for monoatomic ideal gas. Thus,

$$A = 2 \times 12.47 \times (-100) = -2494 \; J = -2.494 \; kJ.$$

Remark: adiabatic expansion is fulfilled at the expense of internal energy of system. Consequently temperature decreases, $\Delta T < 0$ and work is fulfilled in this process: $A < 0$.

Problem № 1.12

64 g oxygen is compressed adiabatically and reversibly at 25^0C from 1 atm until 2 atm. Assume, that $\bar{c_p} = 29.33 \; J / K \times mol$ for oxygen and determine $A, \Delta U, T_2$ and V_2 of process.

Solution:

The following expression is used for calculation of work:

$$A = \frac{nRT_1}{\chi - 1}\left[\left(\frac{P_2}{P_1}\right)^{\frac{\chi-1}{\chi}} - 1\right] \qquad (1.50)$$

where

$$\chi = \frac{\overline{c}_p}{\overline{c}_v} \qquad (1.42)$$

$$\overline{c}_V = \overline{c}_P - R = 29.33 - 8.31 = 21.02 \ J/K\times mol$$

$$\chi = \frac{29.33}{21.02} = 1.395$$

$$\frac{\chi - 1}{\chi} = \frac{0.395}{1.395} = 0.283$$

$$n_{O_2} = \frac{64}{32} = 2 \, mol$$

Let us introduce the obtained values into expression (1.50):

$$A = \frac{2\times 8.31\times 298}{0.395}\left[\left(\frac{2}{1}\right)^{0.283} - 1\right] = 12538.632(1.217-1) = 2717{,}122 \ J$$

$$\Delta U = A = 2.717 \ kJ.$$

Let us use expression (1.40) for determination of T_2:

$$A = n\overline{c}_V (T_2 - T_1)$$

We equalize the right–hand side of expression (1.40) to the value of A, calculated by us:

$$A_{O_2} = n\overline{c}_V (T_2 - T_1) = 2717.122 \ J$$

$$T_2 - T_1 = \frac{2717.122}{2 \times 21.02} = 64.6^0 \text{ K}$$

$$T_2 = 298 + 64.6 = 362.6 \text{ K}$$

The value of V_2 will be calculated from the ideal gas equation:

$$V_2 = \frac{nRT_2}{P_2} = \frac{2 \times 0.082 \times 362.6}{2} = 29.74 \text{ } L$$

Problem № 1.13

By what value will change the work calculated in previous problem, if 2 mole of neon is compressed adiabatic and reversible in the same temperature range?

What value will take P_2 and V_2?

Take into account, that neon is monoatomic gas and its $\bar{c}_V = 12.47$ $J / K \times mol$.

Solution;

The following expression is used for reversible adiabatic compression:

$$A = n\bar{c}_V (T_2 - T_1) \qquad (1.40)$$

Let us calculate work necessary for the compression of neon by using obtained in problem № 1.12 value of ΔT:

$$A_{Ne} = n\bar{c}_V \Delta T = 2 \times 12.47 \times 64.6 = 1612 \text{ } J$$

$$A_{O_2} - A_{Ne} = 2717 - 1612 = 1105 \text{ } J.$$

Equation of adiabate (1.47) is used for estimation of the value of P_2:

$$T^{\chi} P^{1-\chi} = const$$
$$T_1^{\chi} P_1^{1-\chi} = T_2^{\chi} P_2^{1-\chi}$$

$$\left(\frac{P_2}{P_1}\right)^{1-\chi} = \left(\frac{T_1}{T_2}\right)^{\chi} \tag{a}$$

$$\chi = \frac{\overline{c}_p}{\overline{c}_v} = \frac{\overline{c}_v + R}{\overline{c}_v} = \frac{12.47 + 8.31}{12.47} = \frac{20.78}{12.47} = 1.666$$

$$1 - \chi = 1 - 1.666 = -0.666.$$

Let introduce the obtained results into expression (a):

$$\left(\frac{P_2}{1}\right)^{-0.666} = \left(\frac{298}{362.6}\right)^{1.666}$$

$$P_2^{-0.666} = 0.82^{1.666}$$

$$-0.666 \lg P_2 = 1.666 \lg 0.82$$

$$-0.666 \lg P_2 = -0.143$$

$$\lg P_2 = \frac{0.143}{0.666} = 0.215$$

$$P_2 = 10^{0.215} = 1.64 \; atm$$

Let us calculate the value of V_2:

$$V_2 = \frac{nRT_2}{P_2} = \frac{2 \times 0.082 \times 362.63}{1.64} = 36.26 \; L$$

Problem № 1.14

2 moles of neon present at room temperature and under 1 atm pressure. The same amount of work is fulfilled at reversible adiabatic compression of this system, as at compression of 2 moles of oxygen (see problem № 1.12).
    Determine the values of T_2, V_2, P_2 after compression.

Solution:

According to expression (1.40) and condition of problem,

$$A = n\bar{c}_V (T_2 - T_1) = 2717\, J$$

Hence,

$$T_2 - T_1 = \frac{2717}{2 \times 12.47} = 108.9^0\, K$$

$$T_2 = T_1 + 108.9 = 298 + 108.9 = 406.9\, K$$

Let us determine the value of V_2 from the (1.46) equation of adiabate:

$$TV^{\chi-1} = const$$

$$T_1 V_1^{\chi-1} = T_2 V_2^{\chi-1}$$

$$\left(\frac{V_2}{V_1}\right)^{\chi-1} = \frac{T_1}{T_2}$$

$$\chi = \frac{\bar{c}_p}{\bar{c}_v} = \frac{20.78}{12.47} = 1.666$$

$$\chi - 1 = 1.666 - 1 = 0.666$$

$$V_1 = \frac{nRT}{P_1} = \frac{2 \times 0.082 \times 298}{1} = 48.87\, L$$

$$\frac{298}{406.9} = \left(\frac{V_2}{48.87}\right)^{0.666}$$

$$0.732 = \left(\frac{V_2}{48.87}\right)^{0.666}$$

$$\lg 0.732 = 0.666 \lg \left(\frac{V_2}{48.87} \right)$$

$$-0.135 = 0.666 \lg \left(\frac{V_2}{48.87} \right)$$

$$-0.203 = \lg \left(\frac{V_2}{48.87} \right)$$

$$\frac{V_2}{48.87} = 10^{-0.2} = 0.63$$

$$V_2 = 48.87 \times 0.63 = 30.61 \; L$$

$$P_2 = \frac{nRT_2}{V_2} = \frac{2 \times 0.082 \times 406.9}{30.61} = 2.18 \; atm$$

Let us tabulate the results of problems №№ 1.12÷1.14.

Table 1.2. The characteristic values of adiabatic compression

of mono– and twoatomic gases.

(n = 2 mole, $T_1 = 298$ K, $P_1 = 1$ atm, $V_1 = 48{,}87$ L)

Substance	C_V, J / K×mol	ΔT, K	ΔV, L	P_2, atm	A, J	ΔU, J
O_2	21.02	64.6	– 19.1	2	2717	2717
Ne	12.47	64.6	– 12.6	1.64	1612	1612
Ne	12.47	108.9	– 18.3	2.18	2717	2717

As is shown from Table 1.2, less energy expends on the compression of monoatomic gas (Ne), than it is required for the compression of two–atomic gas (O₂) (see problems №1.12 and 1.13). This is conditioned by difference between their heat capacities. Monoatomic gas possesses with a less heat capacity and therefore it may "assimilate" less energy, than two–atomic gas with a higher value of heat capacity (by the the same change of temperature).

On the other hand, by absorption of the same amount of energy, the temperature of gas with a less heat capacity increases more than temperature of gas, which is characterized with a higher value of heat capacity (compare problems № 1.12 and № 1.14).

Accordingly, the value of heat capacity is determined by two factors: assimilation of energy and capability of change of temperature at this assimilation. Moreover, substance with high value of heat capacity assimilates a large amount of energy and changes its own temperature little, but substance with a low value of heat capacity assimilates a minute quantity of energy and changes considerably its own temperature.

Problem № 1.15

One mole of CH_4 is expanded adiabatic and reversible until triple volume. The temperature declines from 298 K to 213 K.

Calculate the values of A, ΔU and ΔH.

Solution:

Fulfilled work is:

$$A = \frac{nR(T_2 - T_1)}{\chi - 1} \qquad (1.41)$$

Equation of adiabate is used for calculation of ($\chi - 1$):

$$TV^{\chi-1} = const \qquad (1.46)$$

$$T_1 V_1^{\chi-1} = T_2 V_2^{\chi-1}$$

$$\left(\frac{V_2}{V_1}\right)^{\chi-1} = \frac{T_1}{T_2}$$

According to condition of problem, $\dfrac{V_2}{V_1} = 3$. Then

$$3^{\chi-1} = \frac{298}{213} = 1.399$$

$$(\chi - 1)\lg 3 = \lg 1.399$$

$$(\chi - 1) \times 0.477 = 0.146$$

$$\chi - 1 = \frac{0.146}{0.477} = 0.306$$

Let us introduce the values of $(\chi - 1)$ and other quantities into expression (1.41):

$$A = \frac{nR(T_2 - T_1)}{\chi - 1} = \frac{1 \times 8.31(213 - 298)}{0.306} = -2308\ J$$

$$\Delta U = A = -2308\ J$$

$\overline{c}_V$ of methane may be determined by two ways:

a) by using of expression (1.43):

$$\overline{c}_V = \frac{R}{\chi - 1} = \frac{8.31}{0.306} = 27.16\ J/K \times mol$$

b) by using of expression (1.40):

$$A = n\overline{c}_V(T_2 - T_1)$$

Hence,

$$\overline{c}_v = \frac{A}{n(T_2 - T_1)} = \frac{-2308}{1(213 - 298)} = \frac{-2308}{-85} = 27.15 \ J/K \times mol$$

$$\overline{c}_p = \overline{c}_v + R = 27.15 + 8.31 = 35.47 \ J/K \times mol$$

$$\Delta H = n\overline{c}_p(T_2 - T_1) = 1 \times 35.47(213 - 298) = 35.47 \times (-85) = -3014 \ J$$

Problem № 1.16
Pressure of 1 mole helium at ambient temperature is equal to 2 atm. Pressure of a gas reduces to 1 atm after adiabatic expansion.
 Determine the values of T_2, V_2, A_{exp}, ΔU and ΔH at reversible and irreversible adiabatic expansion.

Solution:
Find in handbook for He:
$\overline{c}_v = 12.47 \ J/K \times mol$ and $\overline{c}_p = 20.78 \ J/K \times mol$.
 1) Reversible adiabatic expansion.
For calculation of T_2 use the expression (1.47):

$$T^\chi P^{1-\chi} = const$$

$$T_1^\chi P_1^{1-\chi} = T_2^\chi P_2^{1-\chi}$$

$$\left(\frac{P_1}{P_2}\right)^{1-\chi} = \left(\frac{T_2}{T_1}\right)^\chi \qquad (a)$$

$$\chi = \frac{\overline{c}_p}{\overline{c}_v} = \frac{20,78}{12,47} = 1,666$$

$$1 - \chi = -0,666$$

According to conditions of problem, $T_1 = 298 \ K$, $P_1 = 2 \ atm$, $P_2 = 1 \ atm$.

Introduce these values in expression (a):

$$\left(\frac{2}{1}\right)^{-0.666} = \left(\frac{T_2}{T_1}\right)^{1.666}$$

$$0.666 \lg 2 = 1.666 \lg \frac{T_2}{T_1}$$

$$\lg \frac{T_2}{T_1} = -\frac{0.666}{1.666} \lg 2 = -0.4 \times 0.3 = -0.12$$

$$\frac{T_2}{T_1} = 10^{-0.12} = 0.758$$

$$T_2 = T_1 \times 0.758 = 298 \times 0.758 = 225.9 \text{ K}$$

From the ideal gas equation

$$V_2 = \frac{nRT_2}{P_2} = \frac{1 \times 0.082 \times 225.9}{1} = 18.52 \, L$$

Work of adiabatic expansion is equal:

$$A = n\bar{c}_v(T_2 - T_1) = 1 \times 12.47(225.9 - 298) = 12.47 \times (-72.1) = -899.087 J$$

$$\Delta U = A = -899.087 J$$

For calculation of ΔH use expression (1.44):

$$\Delta H = \Delta U + nR\Delta T = -899.087 + 1 \times 8.31 (-72.1) =$$

$$= -899.087 - 599.151 = -1498.238 \, J = - = 1.498 \, kJ$$

2) Irreversible adiabatic expansion opposite constant external pressure.

In this case work is expressed in the following way:

$$A = -P_{ex}\Delta V \qquad (1.8)$$

On the other hand, in adiabatic process (in both reversible and irreversible):

$$A = n\bar{c}_v(T_2 - T_1) \qquad (1.40)$$

Equalize these two expressions:

$$-P_{ex}\Delta V = n\bar{c}_v\Delta T.$$

Hence,

$$\frac{\Delta T}{\Delta V} = -\frac{P_{ex}}{nc_v} = -\frac{1}{1\times 0.123} = -8.13\,\frac{K}{L} \qquad (b)$$

$$(\bar{c}_v = 12.47\,J/K\times mol = 12.47\times 9{,}867\times 10^{-3} = 0.123\,\frac{L\times atm}{K\times mol}\,)$$

Determine the value of V_1 from the ideal gas equation:

$$V_1 = \frac{nRT_1}{P_1} = \frac{1\times 0.082\times 298}{2} = 12.22\,L$$

Depict V_2 by using T_2:

$$P_1V_1 = nRT_1; \qquad P_2V_2 = nRT_2$$

$$\frac{P_2V_2}{P_1V_1} = \frac{T_2}{T_1}$$

According to conditions of problem, $\dfrac{P_2}{P_1} = \dfrac{1}{2}$. Then $\dfrac{V_2}{2V_1} = \dfrac{T_2}{T_1}$ and

$$V_2 = 2V_1 \frac{T_2}{T_1} = \frac{2 \times 12.22}{298} T_2 = 0{,}082 T_2 \qquad (c)$$

Transform expression (b):

$$\frac{\Delta T}{\Delta V} = \frac{T_2 - T_1}{V_2 - V_1} = -8.13$$

$$T_2 - T_1 = -8.13\,(V_2 - V_1) = -8.13\,V_2 + 8.13\,V_1$$

$$T_2 + 8.13\,V_2 = T_1 + 8.13\,V_2 \qquad (d)$$

Introduce the numerical values of T_1 and V_1 and the value of V_2 from expression (c) in the expression (d):

$$T_2 + 8.13 \times 0.082\,T_2 = 298 + 8.13 \times 12.22$$

$$T_2 + 0.66\,T_2 = 298 + 99.35 = 397.35 \text{ K}$$

$$1.66 T_2 = 397.35 \text{ K}$$

$$T_2 = \frac{397.349}{1.66} = 238.50 \ K$$

Determine the value of V_2 from the ideal gas equation:

$$V_2 = \frac{nRT_2}{P_2} = \frac{1 \times 0.082 \times 238.5}{1} = 19.56 \ L$$

For calculation of fulfilled work use expression (1.40):

$$A = n\overline{c}_v (T_2 - T_1) = 1 \times 12.47(238.5 - 298) = 12.47 \times (-59.5) = -741.965 \ J$$

$$\Delta U = A = -741.965 \, J$$

According to expression (1.44):

$$\Delta H = \overline{nc}_p (T_2 - T_1) = 1 \times 20.78 \times (-59.5) = -1236.41 \, J$$

Table 1.3. The characteristic values of adiabatic expansion.

(1 mole He; $T_1 = 298$ K; $V_1 = 12.22$ L; $P_1 = 2$ atm; $P_2 = 1$ atm)

Type of process	T_2, K	ΔT, K	V_2, L	ΔV, L	A, J	ΔU, J	ΔH, J
Reversible	225.9	-72.1	18.52	6.3	-899	-899	-1498
Irreversible	238.5	-59.5	19.56	7.3	-742	-742	-1239

Remark: At reversible adiabatic expansion as well as at irreversible one:

$$\Delta U = \overline{nc}_v \Delta T \qquad (1.40)$$

U represents the state function. Hence, if the initial and final stages are the same in both (reversible and irreversible) processes, then $\Delta U_{rev.} = \Delta U_{irrev.}$.

But at adiabatic expansion even in the case of the same initial conditions $\Delta U_{rev.} \neq \Delta U_{irrev.}$. This is caused by fact that $\Delta T_{rev.} \neq \Delta T_{irrev.}$, *i.e.*

$T_{2\,rev.} \neq T_{2\,irrev.}$ The system does not transfer from the same initial conditions into the same final state at reversible and irreversible adiabatic expansions. In the irreversible process T_2 and V_2 exceed, but fulfilled work is less, than in the reversible one (see also problem N 3.23).

Problem № 1.17

Volume of 5 mole of argon presented at 25^0C and under 1 atm pressure increases four– fold as a results of expansions:1) isothermal reversible,

2) adiabatic reversible and 3) in a vacuum.
   Calculate the values of T_2 , P_2 , A, ΔU for each process.

Solution:

   Following condition of problem, first and second processes proceed reversible. As for third process, it proceeds irreversible, since expansion occurs immediately and $P_{ex} = 0$..
   1) Isothermal reversible expansion ($T_1 = T_2 = const$).
   In condition of n, $T = const$, according to ideal gas equation $PV = const$
, i.e. $P_1V_1 = P_2V_2$. Hence, $P_2 = P_1 \dfrac{V_1}{V_2}$. In accordance with condition of

problem, $\dfrac{V_1}{V_2} = \dfrac{1}{4}$. Thus,

$$P_2 = 1 \times \frac{1}{4} = \frac{1}{4} \, atm.$$

Fulfilled work is:

$$A = -nRT \ln \frac{V_2}{V_1} \tag{1.36}$$

$$A = -5 \times 8.31 \times 298 \ln 4 = -17164 \, J = -17.165 \, kJ$$

In isothermal process:

$$\Delta U = q + A = 0 \tag{1.33}$$

and

$$q = -A = 17.165 \, kJ.$$

2) Adiabatic reversible expansion
Let us use the equation of adiabate for detemination of P_2.

$$PV^{\chi} = const \tag{1.45}$$

$$P_1 V_1^\chi = P_2 V_2^\chi$$

$$P_2 = P_1 \left(\frac{V_1}{V_2} \right)^\chi$$

$$\chi = \frac{\overline{c_p}}{\overline{c_v}} = \frac{20.78}{12.47} = 1.666$$

$$P_2 = 1 \times 0.25^{1.666} = 0.1 \, atm$$

T_2 may be calculated from equation of adiabate (1.46):

$$TV^{\chi-1} = const$$

$$T_1 V_1^{\chi-1} = T_2 V_2^{\chi-1}$$

$$T_2 = T_1 \left(\frac{V_1}{V_2} \right)^{\chi-1} = 298 \times 0.25^{0.666} = 118.4 \text{ K}$$

Work of reversible adiabatic expansion is:

$$A = \frac{nR}{\chi - 1} (T_2 - T_1) \tag{1.41}$$

$$A = \frac{5 \times 8.31}{0.66} (118.4 - 298) = 62.387 \times (-179.6) = -11205 \, J = -11.205 \, kJ$$

In the adiabatic process $q = 0$. Therefore

$$\Delta U = A = -11.205 \, kJ$$

3) Immediate expansion in a vacuum.

Since expansion proceeds in a vacuum, then $P_{ex} = 0$ and

$$A_{irrev.} = -P_{ex}\Delta V = 0$$

Heat transfer does not take place between system and surroundings because of instant expansion: $q = 0$ and $\Delta U = q + A = 0$.

Internal energy of ideal gas is temperature function only. Therefore $\Delta U = 0$ means that $\Delta T = 0$.

Thus, instant expansion of gas in a vacuum represents both isothermal and adiabatic process simultaneously. The work is not fulfilled, heat transfer is not occurred, internal energy and temperature are not changed in this process. The change of volume and pressure take place only.

Let us determine the value of P_2 from ideal gas equation. At $n, T = const$:

$$PV = const$$

$$P_1V_1 = P_2V_2$$

$$P_2 = P_1 \frac{V_1}{V_2} = 1 \times \frac{1}{4} = 0.25 \ atm$$

Let us tabulate the obtained results (Table 1.4).
It is shown from Table (1.4):

1) $|A_{isoth.rev.}| > |A_{adiab.rev.}|$.

$\Delta T = 0$ in isothermal process, but $\Delta T < 0$ in the adiabatic expansion. Therefore $T_{2 \ isoth.} > T_{2 \ adiab.}$ in the same initial conditions. Hence, $P_{2 \ isoth.} > P_{2 \ adiab}$ This ratio of pressures is characteristic not only for final state, but also for the whole process. Work of reversible expansion is: $A_{exp} = -\int\limits_{V_1}^{V_2} PdV$; since P is higher in isothermal process, than in adiabatic one, then $|A_{isoth.}| > |A_{adiab}|$.

2) $|A_{rever.}| > |A_{irrever.}|$.

The work of expansion takes place in reversible isothermal and adiabatic processes. But work is not fulfilled at all in most irreversible process (expansion in a vacuum), proceeded in the same initial and final conditions. The reason lies in the fact that $P_{ex} = 0$ at the expansion in a vacuum.

Table 1.4. Characteristic values of expansion of various kind.

(5 mole Ar; $T_1 = 298$ K; $P_1 = 1$ atm; $V_2/V_1 = 4$)

Kind of process	T_2, K	P_2, atm	A, kJ	q, kJ	ΔU, kJ
Isothermal reversible	298	0.25	$- 17.165$	17.165	0
Adiabatic reversible	118	0.1	$- 11.205$	0	$- 11.205$
In a vacuum irreversible (isothermal--adiabatic)	298	0.25	0	0	0

Problem № 1.18

Is it possible and what work is needed for returning the system and surroundings to the initial state after expansions: 1) isothermal reversible, 2)adiabatic reversible, 3) in a vacuum?

Use the results of problem № 1.17.

Solution:

1) System fulfils -17.165 *kJ* work on surroundings by isothermal reversible four–fold expansion. The pressure of system declines from 1 to 0.25 *atm* (see problem № 1.17).

In order to return the system to its initial state, surroundings must compress the system isothermal and reversible until initial volume of system is achieved. As a result of this process pressure will increase four–fold and reach the initial value. In conditions $n,T = const$, $PV = const$, i.e. $P_1 V_1 = P_2 V_2$ and

$$P_2 = P_1 \frac{V_1}{V_2} = 0.25 \frac{4}{1} = 1 \ atm$$

That energy, which is transferred from system to surroundings at isothermal reversible expansion, is necessary to be transferred from surroundings to the system in order to return surroundings to the initial state. To do this, surroundings must compress system isothermal and reversible. The work fulfilled by surroundings in this process is:

$$A_{revers.} = - nRT \ln \frac{V_2}{V_1} = - 5 \times 8.31 \times 298 \ln \frac{1}{4} = 17165 \ J = 17.165 \ kJ.$$

Thus, after isothermal reversible expansion it is possible to return system and surroundings to the initial state by means of isothermal reversible compression .

It should be mentioned that there is no necessary to carry out the reversible compression for returning the system in the initial state, but it is essential for surroundings. Otherwise surroundings will expend more work on irreversible compression, than it is fulfilled at reversible expansion of system. In result, a certain change will remain in surroundings and it will not return to its initial state.

2) The system has fulfilled -11.205 *kJ* work on surroundings at reversible adiabatic expansion. This results in four–fold increase of system's volume and decrease of pressure from 1 to 0.1 *atm*; temperature is changed from 298 K to 118 K.

In order to return the system and surroundings in the initial state, it is necessary to realize reversible adiabatic compression of system, by which thermodynamic parameters of system restore their initial values, and surroundings will fulfill 11,205 kJ work.

Work, fulfilled by surroundings at adiabatic compression of system from 118 K to 298 K is:

$$A_{rev.\ adiab.} = n\bar{c}_v(T_2 - T_1) = 5 \times 12.47(298 - 118) = 11223\ J = 11.223\ kJ.$$

Pressure of system changes as a result of compression. For determination P_2 we use equation of adiabate (1.47):

$$T^\chi P^{1-\chi} = const$$

Hence

$$\left(\frac{T_2}{T_1}\right)^\chi = \left(\frac{P_1}{P_2}\right)^{1-\chi}$$

$$\chi = \frac{\bar{c}_p}{\bar{c}_v} = \frac{20.78}{12.47} = 1.666$$

$$1 - \chi = -0.666$$

$$\left(\frac{298}{118}\right)^{1.666} = \left(\frac{0.1}{P_2}\right)^{-0.666}$$

$$2.525^{\ 1.666} = \left(\frac{P_2}{0.1}\right)^{0.666}$$

$$4.679 = (10P_2)^{0.666}$$

$$\lg 4.679 = 0.666\ \lg 10P_2$$

$$0.67 = 0.67\ \lg 10P_2$$

$$\lg 10\, P_2 = 1$$

$$10\, P_2 = 10$$

$$P_2 = 1 \text{ atm}$$

The value of V_2/V_1 may be determined from equation of adiabate (1.46):

$$TV^{\chi-1} = const$$

Hence,

$$\frac{T_2}{T_1} = \left(\frac{V_1}{V_2}\right)^{\chi-1}$$

$$\frac{298}{118} = \left(\frac{V_1}{V_2}\right)^{0.666}$$

$$2.525 = \left(\frac{V_1}{V_2}\right)^{0.666}$$

$$\lg 2.525 = 0.666 \lg \frac{V_1}{V_2}$$

$$0.402 = 0.666 \lg \frac{V_1}{V_2}$$

$$\lg \frac{V_1}{V_2} = 0.604$$

$$\frac{V_1}{V_2} = 10^{0.604} = 4$$

Thus, if surroundings expend so much work on equilibrium compression of system as is fulfilled by system on surroundings at equilibriom expansion, then both surroundings and system will return to their initial state.

3) Expansion in a vacuum is a most irreversible process, at which work is not fulfilled at all:

$$A_{exp} = -P_{ex}\Delta V = 0 \qquad (P_{ex} = 0)$$

But it is necessary to fulfil work for returning (for compression) system to the initial state. Because expansion in a vacuum proceeds isothermal, therefore let us return system to its initial state by isothermal reversible compression. The work, fulfilled by surroundings in this process is:

$$A_{compr.\ rever.} = -nRT \ln \frac{V_2}{V_1} = -5 \times 8.31 \times 298 \ln \frac{1}{4} = 17165\ J = 17.165\ kJ.$$

Thus, returning of system expanded irreversible in a vacuum to the initial state is possible via reversible isothermal compression. But returning of surroundings to the initial state is impossible, it will have a shortage of energy 17.165 *kJ*. If the system is returned to the initial state via irreversible isothermal compression, then surroundings removes further from the initial state; it will have to fulfill more than 17.165 *kJ* work, since $A_{compr.\ irrev.} > A_{compr.\ revers.}$.

It may be mentioned at the end, that returning of system and surroundings to the initial state is possible only after equilibrium expansion. A certain change remains in system or in surroundings,either in both after nonequilibrium (irreversible) expansion.

2. Thermochemistry

Theoretical Part

Thermochemistry represents a part of chemical thermodynamics, which provides determination of heats for chemical, physical and physico–chemical processes (chemical reaction, phase transfer, solution, dilution, etc); The changes of internal energy (U) and enthalpy (H) in the system may be estimated on the basis of heat of process.

Heat of reaction

Let us consider chemical reaction proceeded in thermodynamic system:

$$\nu_A A + \nu_B B + \ldots \rightarrow \nu_M M + \nu_N N + \ldots \qquad (I)$$

where ν_i is stoichiometric coefficient of i–th compound.

Internal energy of system is a function of temperature, volume and composition. The distinction between initial and final values of internal energy at constant volume and temperature ($V, T = const$) depends on changes in chemical composition of system only:

$$\Delta U_{syst.} = U_2 - U_1 = \sum_{products} \nu_i U_i - \sum_{reactants} \nu_i U_i \qquad (2.1)$$

where U_i is internal energy of i–th compound at the given conditions.

The difference between the internal energies (or enthalpies) of reactants and products represents heat of reaction Q.

This distinction reveals itself at isothermal conditions by heat transfer with the surroundings. Isothermal heat is determined in conditions $V = const$ or $P = const$:

$$Q_V = \Delta U_{syst.} \qquad (2.2)$$

$$(V = const, \quad T_1 = T_2)$$

The decrease of internal energy ($\Delta U < 0$) of system as a result of reaction means, that the internal energy of initial state of system (i.e. energy of reactants) exceeds the energy of final state (i.e. energy of products). The difference between them releases in the system as heat. This heat must be transferred to the surroundings in order to protect isothermality of the process (otherwise temperature of system increases: $T_2 > T_1$ and process will not be isothermal any more). Heat, transferred by system to surroundings is negative:

$$Q_V = \Delta U < 0 \qquad \text{(reaction is exothermic)}$$

If internal energy of system increases as a result of the process ($\Delta U > 0$), then the system receives an energy, necessary for the reaction from the surroundings in the form of a heat. (If system performs the process via its "own forces", then temperature decreases: $T_2 < T_1$ and the condition of isothermality is disturbed). Heat received by system is positive:

$$Q_V = \Delta U > 0 \qquad \text{(reaction is endothermic)}$$

The distinguished form of conditions recording ($V = const, \quad T_1 = T_2$) in expression (2.2) is not occasional. The volume must remain constant during a whole process. Otherwise the expansion work will fulfil and Q would not be egual to ΔU any more. As regards the temperature, it may be varied at the proceeding of the reaction. But temperatures must be the same ($T_1 = T_2$) at the beginning and the end of process (i.e. temperatures of reactants and products).

Let us assume a contrast and suggest that temperature of a system rises as a result of exothermic reaction. Thus products will present at higher temperature, than reactants before beginning of reaction: $T_{prod.} > T_{react.}$ (i.e.

$T_2 > T_1$). Therefore not only internal energy corresponded to T_1 temperature is accumulated in the products, but also energy of heating from T_1 to T_2. The latter represents a part of heat released in the system as a result of reaction. If it is neglected, a precise estimation of heat is impossible. In order to determine heat exactly, the cooling of products to T_1 temperature and taking into account the heat released by cooling is necessary.

The enthalpy is considered instead of internal energy by carrying out of process at isobaric conditions and at given temperature. Enthalpy represents a function of temperature, pressure and composition of system. If reaction is carried out at constant pressure and temperature, then

$$\Delta H_{syst.} = H_2 - H_1 = \sum_{products} v_i H_i - \sum_{reactants} v_i H_i \qquad (2.3)$$

where H_i is enthalpy of i–th compound in the given conditions.

$$Q_P = \Delta H_{syst.} \qquad (2.4)$$

$$(P = const, \quad T_1 = T_2)$$

$Q_P = \Delta H < 0$ in exothermic reaction, but $Q_P = \Delta H > 0$ in endothermic one.

Thus, heat transferred to surroundings or received from surroundings by the system in isochoric or isobaric and $T_1 = T_2$ conditions is called isothermal heat of reaction. The following requirements must be executed at the carrying out of process: a) temperatures of reactants and products must be the same; b) work of all kind must be excluded at proceeding of reaction, except the work of expansion.

Heat of reaction is measured by using of calorimeter. In order to protect an isothermality of the process, transfer of heat to calorimeter or absorption of heat from calorimeter must occure as soon as possible. (Calorimeter represents surroundings in this case).

The determination of heat is performed by change of temperature of calorimeter:

$$Q_V = - C_{V cal.} \times \Delta T_{cal.} \qquad (2.5)$$

where $C_{cal.}$ is the heat capacity of calorimeter.

Heat of reaction may be determined in adiabatic conditions also, i.e. when heat transfer does not take place between system and surroundings: $q_{adiab.} = 0$.

Let us represent the first law of thermodynamics as follows:

$$\Delta U_{syst.} = \Delta_e U + \Delta_i U \qquad (2.6)$$

where $\Delta_e U$ is the change of internal energy of system induced by interaction with surroundings (heat transfer and work):

$$\Delta_e U = q + A \qquad (2.7)$$

$\Delta_i U$ represents the change of internal energy, caused by processes inside the system (without interaction with surroundings). According of law of energy constancy, the formation or annihilation of energy do not occur in the system. The interconversion of various kinds of energy may proceed only. Therefore

$$\Delta_i U = 0 \qquad (2.8)$$

1) Adiabatic–isochoric process ($q = 0$, $V = const$).

Work of expansion is not fulfilled in these conditions: $A = 0$ and heat transfer with surroundings is not occurred: $q = 0$. According to expression (2.7),

$$\Delta_e U = q + A = 0.$$

Thus, internal energy of system does not change in adiabatic–isochoric process:

$$\Delta_i U = 0, \qquad \Delta_e U = 0$$

and

$$\Delta U_{syst.} = \Delta_e U + \Delta_i U = 0.$$

Notwithstanding heat of chemical reaction proceeded in the system does not equal to zero. It is equal to the difference in energies of products and reactants at initial T_1 temperature:

$$Q_V^* = \sum_{products} v_i U_i - \sum_{reactants} v_i U_i \qquad (2.9)$$

$$(q = 0 \qquad \text{and} \qquad V = const)$$

a) Internal energy of reactants exceeds internal energy of products:

$$\sum_{reactants} v_i U_i > \sum_{products} v_i U_i$$

This difference of energies releases inside the system at the proceeding of reaction and represents heat of reaction (Q_V^*) [1]. The transfer of this heat to surroundings is excluded due to the adiabaticity of process. It remains in the system and causes a rising of temperature of the system: $T_2 > T_1$. (A calorimeter with substances inside it represents the thermodynamic system in this case.). Transformation of a part of "chemical" energy, accumulated in reactants takes place into the equivalent amount of "heat" energy in considered process.

$$Q_V^* < 0 \qquad \qquad \text{(reaction is exothermic)}$$

Q_V^* may be determined, if heat capacity $(C_{V\ cal.})$ and temterature change of calorimeter are known:

$$Q_{V\ cal.}^* = - C_{V\ cal.}\ (T_2 - T_1) \qquad (2.5)$$

In this case $T_2 > T_1$ and $Q_V^* < 0$.

b) The internal energy of reactants is less than internal energy of products:

---

[1] Heat of adiabatic–isochoric and adiabatic–isobaric processes are designated by Q_V^* and Q_P^*

$$\sum_{reactants} v_i U_i \ < \ \sum_{products} v_i U_i$$

Heat transfer does not occur between system and surroundings due to adiabaticity of the process. Therefore the system borrows energy, necessary for the carrying out the chemical reaction from its own "margin". Hence, internal energy of system decreases and temperature declines: $T_2 < T_1$. The transformation of a part of "heat" energy of the system into the equivalent amount of "chemical" energy takes place in the given process.

Heat of reaction according to expression (2.9) is:

$$Q_V^* = \sum_{products} v_i U_i \ - \ \sum_{reactants} v_i U_i > 0$$

The reaction is endothermic.

According to expression (2.5),

$$Q_{V\,cal.}^* = - C_{V\,cal.} \,(T_2 - T_1)$$

$$(T_2 < T_1 \quad \text{and} \quad Q_V^* > 0).$$

2) Adiabatic–isobaric process $(q = 0, \quad P = const)$.

As is known, at $P = const$

$$\Delta H_{syst.} = q_p \tag{1.16}$$

$q_p = 0$ in the adiabatic process. Thus, if reaction proceeds in adiabatic–isobaric conditions, then $\Delta H_{syst.} = 0$. In the same time heat of reaction Q_P^* is equal to the difference of enthalpies of products and reactants:

$$Q_P^* = \sum_{products} v_i H_i \ - \ \sum_{reactants} v_i H_i \tag{2.10}$$

$$(q = 0 \quad \text{and} \quad P = const)$$

The following expression is used for the determination of Q_P^*:

$$Q_P^* = = - C_{P\,cal.}\,(T_2 - T_1) \tag{2.11}$$

Enthalpies of reactants exceed enthalpies of products:

$$Q_P^* = \sum_{products} v_i H_i - \sum_{reactants} v_i H_i < 0$$

The reaction is exothermic; the rising of temperature proceeds in the system: $T_2 > T_1$.

b) The sum of enthalpies of reactants is less than sum of enthalpies of products:

$$Q_P^* = \sum_{products} v_i H_i - \sum_{reactants} v_i H_i > 0$$

The reaction is endothermic and $T_2 < T_1$.

Finally, the following statement should be emphasized once again: heat of chemical reaction represents the difference in the internal energies (enthalpies) of products and reactants at the initial temperature. Heat of reaction must not be identified with the change of internal energy (enthalpy) of the system.

The value of heat of reaction in isothermal conditions ($T_1 = T_2$) coincides with the change of internal energy (enthalpy) of system:

$$Q_V = \Delta U_{syst.} \qquad \text{and} \qquad Q_P = \Delta H_{syst.}$$

Such coinciding does not take place in adiabatic conditions ($q = 0$):

$$Q_V^* \neq \Delta U_{syst.} \qquad \Delta U_{syst} = 0$$

$$Q_P^* \neq \Delta H_{syst.} \qquad \Delta H_{syst} = 0$$

Notwithstanding, the value of heat of reaction is the same in isothermal and adiabatic processes:

$$Q_V = Q_V^* \qquad \text{and} \qquad Q_P = Q_P^*$$

Anything abovementioned covers phase transformation, dissolution, dilution, etc.

Isobaric–isothermal heat of reaction will now be considered, since most of the chemical reactions and physico–chemical transformations proceed at constant pressure and temperature. The change of enthalpy of the system will be used for the representation of isobaric heat of reaction:

$$Q_P = \Delta H_{syst.} = H_2 - H_1 \qquad \text{(when } T_1 = T_2)$$

Hess's law

Hess's law represents a particular kind of law of energy constancy for those systems, in which chemical, physical or physico–chemical transformations take place. Its analytical expression is:

$$Q_P = \Delta H \qquad (2.4)$$

$$(P = const, \quad T_1 = T_2)$$

This law indicates, that energy released in the system as a result of chemical reaction is transferred to the surroundings in the form of equivalent amount of heat (exothermic reaction, $Q_P < 0$); in contrast to this, heat received from surroundings is expended on the carrying out a chemical reaction in the system (endothermic reaction, $Q_P > 0$).

Enthalpy is a state function, i.e. its change depends on the initial and final states of system and is independent of way of process:

$$\Delta H = H_2 - H_1$$

Enthalpy is an additive quantity:

$$\Delta H_{chem.reac.} = \Delta H_1 + \Delta H_2 + \ldots + \Delta H_n \qquad (2.12)$$

where ΔH_1, ΔH_2, ΔH_n are the heat values of those stages, into which the considering reaction may be divided.

Finally, Hess's law may be summarized so: heat of chemical reaction is independent of process's way, but depends on the initial and final states of system only. Heat of reaction may be represented as a sum of heats of those processes, into which the reaction of interest may be divided (really or imaginary).

Hess's law is followed by some statements:

1) Heat of reaction is equal to the difference in heats of formation of products and reactants:

$$\Delta H_{chem.\ reaction} = \sum_{prod.} v_j (\Delta H_j)_{form.} - \sum_{react.} v_i (\Delta H_i)_{form.} \qquad (2.13)$$

where v_i and v_j are the stoichiometric coefficients of corresponded compounds in the equation of the reaction; ΔH_i and ΔH_j represent heats of formation (enthalpy) of reactants and products respectively.

Heat of formation of 1 mole of the given compound from elementary substance is called heat of formation of substance. (It is accepted that heat of formation of elementary substance is equal to zero.). $\Delta H_{form.}$ may be either positive or negative quantity. $\Delta H_{form} > 0$ means, that expending of energy is necessary for production of substance. $\Delta H_{form} < 0$ means, that heat is released at the formation of substance.

2) Heat of reaction is equal to the difference in heat of combustion of reactants and products:

$$\Delta H_{chem.reac.} = \sum_{reactant} v_i (\Delta H_i)_{combustion} - \sum_{product} v_j (\Delta H_j)_{combustion} \qquad (2.14)$$

Heat of complete combustion of 1 mole of substance is called heat of combustion. Heat of combustion is always negative quantity: $\Delta H_{comb.} < 0$ i.e. heat is always released at the combustion.

3) Heat of reaction is equal to the difference in energy of broken and formed bonds:

$$\Delta H_{chem.\ react.} = \sum_{reactant} n_i \varepsilon_i - \sum_{product} m_i \varepsilon_i \qquad (2.15)$$

where ε_i is breaking energy of bond of i-th kind, n_i and m_i represent number of bonds of respective kinds.

Heat of reaction, in which the breaking of 1 mole of the given bond occurs, is called breaking energy of bond. The breaking energy of bond is always positive: $\Delta H_{bond\ break.} > 0$. This means, that the expenditure of energy is necessary to break bond of any kind. In contrast to this, energy is always released at the formation of bond. The reaction is endothermic ($\Delta H_{chem.reac.} > 0$), if the sum of energies of bonds broken in the chemical reaction exceeds the sum of energies of formed bonds. Otherwise the reaction is exothermic: $\Delta H_{chem.reac.} < 0$.

The heat of phase transformations (melting, solidification, evaporization, condensation, allotropic transformation) is often used in thermochemistry.

Enthalpy change of 1 mole of substance at the transferring from one phase to the other, is called heat of phase transformation ($\Delta H_{ph.tr.}$). According to Hess's law, $\Delta H_{ph.tr}$ represents the difference in heat formations of substance in two phases. If the following phase transformation occurs:

$$\text{phase (1)} \rightarrow \text{phase (2)},$$

then

$$\Delta H_{ph.tr} = \Delta H_{form.(2)} - \Delta H_{form.(1)} \qquad (2.16)$$

where $\Delta H_{form.(1)}$ and $\Delta H_{form.(2)}$ are heats of formation of substance in (1) and (2) phases respectively.

The standardization of heats is needed for carrying out a thermochemical calculations, i.e. its consideration in the same conditions.

Heats of chemical reaction, phase transformation, dissolution, dilution and other processes in the condition of 1 *atm* pressure (1.013×10^5 *Pa*) and constant temperature (T K) is called standard heat. It is designated as ΔH_T^0. The values of heats, given in handbooks correspond to 1 *atm* pressure and 298 K. Its designation is : ΔH^0_{298}. Dimensions of molar heat are: *J / mol* or *cal / mol* (1 *cal* = 4.184 *J*).

Dependence of heat of reaction on the temperature (Kirchhoff's law)
Heat of most chemical reactions is depended on the temperature. Let us consider scheme of reaction:

$$v_A A + v_B B + ... \rightarrow v_M M + v_N N + ... \qquad (I)$$

Temperature dependence of heat of reaction is represented so:

$$\frac{d\Delta H_{chem.reac.}}{dT} = \Delta C_P \qquad (2.17)$$

where

$$\Delta C_P = (v_M C_{p\,(M)} + v_N C_{p\,(N)}) - (v_A C_{p\,(A)} + v_B C_{p\,(B)}) \qquad (2.18)$$

(2.17) represents an analytical expression of Kirchhoff's law. It indicates, that the dependence of heat of reaction on the temperature is determined by the change of heat capacity of system during process. Let us consider the following cases:

1) $\Delta C_P = 0$,   then $\dfrac{d\Delta H_{chem.reac.}}{dT} = 0$

If heat capacity of system does not change at the proceeding of reaction, then heat of reaction is independent of temperature.

2) $\Delta C_P > 0$,   then $\dfrac{d\Delta H_{chem.reac.}}{dT} > 0$

If heat capacity of products exceeds heat capacity of reactants (heat capacity of system increases at the proceeding of reaction), then the value of heat of reaction with rising of temperature becomes more positive or less negative. In other words, the value of heat of endothermic reaction increases and the value of exothermic reaction decreases with rising a temperature.

3) $\Delta C_P < 0$,   then $\dfrac{d\Delta H_{chem.reac.}}{dT} < 0$

If heat capacity of products is less than heat capacity of reactants (heat capacity of system decreases at the proceeding of reaction), the value of heat of reaction becomes less positive (in endothermic reaction) or more negative (in exothermic reaction) with rising a temperature. So the value of endotermic reaction's heat decreases and the value of exothermic reaction's heat increases with rising a temperature.

Such relationship of ΔC_P with temperature dependence of heat of reaction is explained by temperature dependence of C_P substances', participating in reaction. Temperature dependence of C_P of the individual substance is expressed by power series:

$$C_P = a + b\,T + c\,T^2 + c'\,T^{-2} \tag{2.19}$$

where a, b, c and c' are coefficients.
Hence,

$$\Delta C_P = \Delta a + \Delta b\,T + \Delta c\,T^2 + \Delta c'\,T^{-2} \tag{2.20}$$

The values of C_P of substances are changed differently with the change of temperature. This may cause change of sign of ΔC_P and formation of extremum in relationship: $\Delta H = f(T)$ (see problem № 2.29).

The following form of Kirchhoff's equation is mostly used in the calculations:

$$\Delta H_{T_2} = \Delta H_{T_1} + \int_{T_1}^{T_2} \Delta C_P dT \tag{2.21}$$

Since most data are at 298 K, the following expression is often used:

$$\Delta H_T = \Delta H_{298} + \int_{298_1}^{T} \Delta C_P dT \tag{2.22}$$

Let us introduce equation (2.20) into (2.21):

$$\Delta H_{T_2} = \Delta H_{T_1} + \Delta a\,(T_2 - T_1) + \frac{\Delta b}{2}\,(T_2^2 - T_1^2) + \frac{\Delta c}{3}\,(T_2^3 - T_1^3) -$$

$$- \Delta c'\left(\frac{1}{T_2} - \frac{1}{T_1}\right) \tag{2.23}$$

If temperature range is narrow, the values of average heat capacity may be used:

$$C_{P_{av.}} = \frac{1}{T_2 - T_1} \int_{T_1}^{T_2} C_P dT \qquad (2.24)$$

$$\Delta C_{P_{av.}} = \sum_{prod.} v_j C_{j_{av.}} - \sum_{react.} v_i C_{i_{av.}} \qquad (2.25)$$

where v_i and v_j are stoichiometric coefficients of substances participated in the reaction.

Expression (2.21) obtains the following form in this case:

$$\Delta H_{T_2} = \Delta H_{T_1} + \Delta C_P \Delta T \qquad (2.26)$$

Ratio between ΔH and ΔU.

As is known,

$$\Delta H = \Delta U + P\Delta V + V\Delta P + \Delta P\Delta V = \Delta U + P_2 V_2 - P_1 V_1 \qquad (1.11)$$

In the chemical reaction:

$$\Delta H = \Delta U + \Delta(PV) = \Delta U + (PV)_{products} - (PV)_{reactants} \qquad (2.27)$$

PV of reactants and products in processes, proceeded by participation of solid and liquid substances, differ negligibly and therefore it may be accepted that $\Delta H \approx \Delta U$. But the change of PV must be essentially considered, when gases take part in the reaction. Let us consider each gas as ideal and change their PV with nRT. Then in isothermal process:

$$\Delta(PV) = RT\Delta n_{gas} \qquad (2.28)$$

where

$$\Delta n_{gas} = n_{gas(prod.)} - n_{gas\ (react.)} \qquad (2.29)$$

Finally it is obtained:

$$\Delta H = \Delta U + RT\Delta n_{gas} \qquad (2.30)$$

Problems

Problem № 2.1

Calculate heat of reaction:

$$NH_3 + 5/4\ O_2 \rightarrow NO + 3/2\ H_2O_{(g)} \qquad (I)$$

if the following data are given:

       (1) $\frac{1}{2} N_2 + 3/2\ H_2 \rightarrow NH_3$ $\Delta H_1 = -46190\ J/mol$

       (2) $NO \rightarrow \frac{1}{2} N_2 + \frac{1}{2} O_2$ $\Delta H_2 = -90370\ J/mol$

       (3) $H_2 + \frac{1}{2} O_2 \rightarrow H_2O_{(liq.)}$ $\Delta H_3 = -285840\ J/mol$

       (4) $H_2O_{(g)} \rightarrow H_2O_{(liq.)}$ $\Delta H_4 = -44000\ J/mol$

Solution:

According to one of results of Hess's law, heat of reaction may be represented so:

$$\Delta H_{reaction} = \sum_{prod.} v_j (\Delta H_j)_{form.} - \sum_{react.} v_i (\Delta H_i)_{form.} \qquad (2.13)$$

Then

$$\Delta H_{reaction} = \Delta H_{form.NO} + 3/2\ \Delta H_{form.H_2O(g)} - \Delta H_{form.NH_3} \qquad (II)$$

As is shown from condition of problem,

$$\Delta H_{form.NO} = -\Delta H_2 = 90370 \ J / mol$$

$$\Delta H_{formNH_3} = \Delta H_1 = -46190 \ J / mol$$

$$\Delta H_{formH_2O(g)} = \Delta H_{formH_2O(liq.)} + \Delta H_{evapor.H_2O} = \Delta H_3 - \Delta H_4 =$$

$$= -285840 + 44000 = -241840 \ J / mol$$

Let us introduce these data into expression (II):

$$\Delta H_{reaction} = 90370 + 3/2 \ (-241840) - (-46190) =$$

$$= 90370 - 362760 + 46190 = -226200 \ J/mol$$

Problem № 2.2
The reaction:

$$CaCO_3 \rightarrow CaO + CO_2$$

is given.
Determine the value of its heat at 298 K under conditions of: 1) constant pressure and 2) constant volume.

Solution:

1) According to one of results of Hess's law, heat of reaction in conditions of constant pressure is equal:

$$\Delta H_{reaction} = \sum_{prod.} v_j (\Delta H_j)_{form.} - \sum_{react.} v_i (\Delta H_i)_{form.} \quad (2.13)$$

In our case:

$$\Delta H_{reaction} = \Delta H_{form CO_2} + \Delta H_{form.CaO} - \Delta H_{form.CaCO_3} \quad (I)$$

The standard heat of compounds participating in reaction at 298 K is found in the handbook:

Substance	$\Delta H^0_{298\,form.}$, kJ/mol
CO_2	−393.51
CaO	−635.1
$CaCO_3$	−1206

Let us introduce these data into (I):

$$\Delta H_{reaction} = -393.51 - 635.1 - (-1206) = 177.39\ kJ/mol.$$

2) Heat of reaction in the conditions of constant volume is equal to the change of internal energy (ΔU). As is known,

$$\Delta H = \Delta U + RT\Delta n \qquad (2.30)$$

where Δn is a change of number of moles in gaseous phase.
In the given reaction $\Delta n_{gas} = 1$. Then

$$\Delta U = \Delta H - RT = 177390 - 8.31 \times 298 = 174913.62\ J/mol =$$

$$= 174.914\ kJ/mol$$

Problem № 2.3
Determine standard heat of reaction:

$$CH_4 + CO_2 \rightarrow 2\ CO + 2\ H_2 \qquad (I)$$

following the given data:

1) $CH_4 + 2\ O_2 \rightarrow CO_2 + 2\ H_2O$ $\qquad\qquad$ $\Delta H_1 = -890.31\ kJ/mol$

2) $CO + \tfrac{1}{2}\ O_2 \rightarrow CO_2$ $\qquad\qquad$ $\Delta H_2 = -282.74\ kJ/mol$

3) $H_2 + \tfrac{1}{2}\ O_2 \rightarrow H_2O_{(liq.)}$ $\qquad\qquad$ $\Delta H_3 = -285.04\ kJ/mol$

Solution:

According to one of results of Hess's law:

$$\Delta H_{reaction} = \sum_{reactants} v_i (\Delta H_i)_{combustion} - \sum_{products} v_j (\Delta H_j)_{combustion} \qquad (2.14)$$

ΔH_1, ΔH_2, ΔH_3 represent heat of combustion of substances participating in the reaction (I); (heat of combustion of CO_2 is equal to zero). Hence, it is obtained:

$$\Delta H_{react.} = \Delta H_1 - 2\Delta H_2 - 2\Delta H_3 = -890.31 - 2(-282.74) - 2(-285.84) =$$
$$= -890.31 + 565.48 + 571.68 = 246.85 \ kJ/mol$$

Problem № 2.4

Determine heat of isomerization of cyclopropane into propylene, if heats of combustion of cyclopropane and propylene are equal to −2091 kJ/mol and −2058 kJ/mol respectively.

Solution:

According to Hess's law, heat of reaction $(CH_2)_3 \rightarrow C_3H_6$ may be determined by using heats of combustion of substances, taking part in reaction:

$$\Delta H_{isom.} = (\Delta H_{cyclopr.})_{comb.} - (\Delta H_{propylene})_{comb.} = -2091 - (-2058) = -33 \ kJ/mol$$

Problem № 2.5

Find heat of combustion of benzol, if heat of its formation is +48.99 kJ/mol, but heats of combustion of H_2 and C are equal to −142.92 kJ/mol and −393.51 kJ/mol respectively.

Solution:

Let us imagine, that benzol is formed by following reaction:

$$6 \ C + 6 \ H \rightarrow C_6H_6 \qquad (I)$$

Heat of formation of benzol is equal to heat of reaction (I):

$$\Delta H_{\text{form. benz..}} = \Delta H_I$$

On the other hand, heat of reaction (I) may be represented by using of combustion heats of substances, which take part in the reaction:

$$\Delta H_I = 6 \, \Delta H_{\text{comb. C}} + 6\Delta H_{\text{comb. H}} - \Delta H_{\text{comb.benz.}}$$

Hence

$$\Delta H_{\text{comb. benz.}} = 6 \, \Delta H_{\text{comb. C}} + 6 \, \Delta H_{\text{comb. H}} - \Delta H_{\text{form. benz.}} = 6 \times (-393.51) +$$

$$+ \, 6 \times (-142.92) - 48.99 = -3267.57 \; kJ / mol$$

Problem № 2.6
Find heat of reaction:

$$4NH_3 + 3O_2 \rightarrow 2N_2 + 6H_2O_{(g)} \tag{1}$$

if heats of formation of ammonia and water are equal to –46.11 kJ/mol and –241.82 kJ/mol respectively.

Solution:
According to condition of problem:

$$(2) \; \tfrac{1}{2} \, N_2 + 3/2 \, H_2 \rightarrow NH_3 \qquad\qquad \Delta H_2 = -46.11 \; kJ / mol$$

$$(3) \; H_2 + 1/2 \, O_2 \rightarrow H_2O \qquad\qquad \Delta H_3 = -241.82 \; kJ / mol$$

Let us multiply the stoichiometric coefficients of reaction (2) by four, but coefficients of reaction (3) by six. Correspondingly the values of heat will change:

$$(4) \; 2 \, N_2 + 6 \, H_2 \rightarrow 4 \, NH_3 \qquad\qquad \Delta H_4 = 4 \, \Delta H_2 = -184.44 \; kJ / mol$$

$$(5) \; 6 \, H_2 + 3 \, O_2 \rightarrow 6 \, H_2O \qquad\qquad \Delta H_5 = 6 \, \Delta H_3 = -1450.92 \; kJ / mol$$

Let us substract reaction (4) from the reaction (5). Then it is obtained:

$$6 H_2 + 3 O_2 - 2 N_2 - 6 H_2 \rightarrow 6 H_2O - 4NH_3$$

$$3 O_2 - 2 N_2 \rightarrow 6 H_2O - 4 NH_3$$

$$4 NH_3 + 3 O_2 \rightarrow 2 N_2 + 6 H_2O \qquad (1)$$

Thus, if reaction (4) is substracted from reaction (5), then reaction (1) is obtained. According to Hess's law:

$$\Delta H_1 = \Delta H_5 - \Delta H_4 = -1450.92 - (-184.44) = -1266.48 \; kJ / mol.$$

Problem № 2.7
The values of bond breaking are given:

Bond	ΔH_{bond}, kJ / mol
N—H	391
O=O	498
N≡N	941
H—OH	498
O—H	429

Calculate standard heat of reaction:

$$4NH_3 + 3O_2 \rightarrow 2N_2 + 6H_2O \qquad (I)$$

at 298 K.

Solution:
Heat of reaction may be expressed by using of energies of broken and formed bonds:

$$\Delta H_{reaction} = \sum_{reactant} n_i \varepsilon_i - \sum_{product} m_i \varepsilon_i \qquad (2.15)$$

where ε_i is energy of bond of i–th kind, n_i and m_i represent number of bonds of respective type in reactants and products.

For reaction (I):

$$\Delta H_{reaction} = 12\,\varepsilon_{N\text{-}H} + 3\,\varepsilon_{O=O} - 2\,\varepsilon_{N\equiv N} - 6\,\varepsilon_{H\text{-}OH} - 6\,\varepsilon_{O\text{-}H} =$$

$$= 12 \times 391 + 3 \times 498 - 2 \times 941 - 6 \times 498 - 6 \times 429 =$$

$$= 4692 + 1494 - 1882 - 2988 - 2574 = -1258\ kJ\,/\,mol$$

Heat of the same reaction in problems № 2.6 and № 2.7 is determined by using of two different way. The results are approximately identical (difference makes up ~ 0.7%).

Problem № 2.8

Energy of H–H bond is 436 kJ / mol and energy of O=O bond is equal to 498 kJ / mol.

What is the heat of atomization (decomposition into atoms) of hydrogen peroxide, if heat of formation of this compound is – 187.8 kJ / mol?

Solution:

The process of atomization of hydrogen peroxide may be expressed so:

$$H_2O_2 \rightarrow 2O + 2H \tag{1}$$

Imagine, that reaction (1) proceeds step–wise:

$H_2O_2 \rightarrow O_2 + H_2$	(2)
$H_2 \rightarrow 2\,H$	(3)
$O_2 \rightarrow 2\,O$	(4)

$$H_2O_2 \rightarrow 2O + 2H \qquad\qquad (1)$$

Following Hess's law:

$$\Delta H_1 = \Delta H_2 + \Delta H_3 + \Delta H_4$$

According to condition of problem,

$$\Delta H_2 = -\Delta H_{form. H_2O_2} = -(-187.8) = 187.8 \ kJ/mol$$

$$\Delta H_3 = 436 \ kJ/mol \quad \text{and} \quad \Delta H_4 = 498 \ kJ/mol$$

Then

$$\Delta H_1 = \Delta H_2 + \Delta H_3 + \Delta H_4 = 187.8 + 436 + 498 = 1121.8 \ kJ/mol.$$

Problem № 2.9

Calculate heat of atomization of hydrogen peroxide by other mode in contrast to problem № 2.8, if energies of O–O and O–H bonds are equal to 139 kJ / mol and 926 kJ / mol respectively.

Solution:

As is mentioned in problem № 2.8, process of atomization of hydrogen peroxide is represented so:

$$H_2O_2 \rightarrow 2H + 2O \tag{1}$$

Let ue refer to Hess'a law again and carry out the process (1) in a manner of steps, which differs from the way, considered in problem № 2.8. Imagine, that firstly O—O bond breaks in hydrogen peroxide and then O–H bond breaks in the obtained OH radical:

$$H_2O_2 \rightarrow 2 \ OH$$
$$+ \tag{2}$$
$$2 \ OH \rightarrow 2 \ O + 2 \ H \tag{3}$$

$$H_2O_2 \rightarrow 2 \ H + 2 \ O \tag{1}$$

According to Hess's law:

$$\Delta H_1 = \Delta H_2 + \Delta H_3 = 139 + 926 = 1065 \ kJ/mol$$

Thus, the comparison of problems №№ 2.8 and 2.9 reveals, that approximately identical results are obtained by carrying out just the same process in two different ways.

Problem № 2.10

The reaction proceeds at constant temperature (298 K) and constant pressure (1 atm). What are the values of ΔH, ΔU and heat of reaction:

$$NH_4Cl_{(solid)} \rightarrow NH_3 + HCl$$

Does the system expand or compress at the proceeding of reaction?

Solution:

The change of number of moles in the gaseous phase is:

$$\Delta n_{gas} = 1 + 1 - 0 = 2 > 0$$

Thus, the system expands at the proceeding of reaction.
Let us find the value of reaction's ΔH by using of a handbook:

$$\Delta H_{reaction} = \Delta H_{form.NH_3(gas)} + \Delta H_{form.HCl(gas)} - \Delta H_{form.NH_4Cl(solid)} =$$
$$= -46.11 - 91.83 - (-314.4) = 176.46 \ kJ/mol.$$

The following relationship exists between ΔH and ΔU:

$$\Delta H = \Delta U + RT\Delta n_{gas} \qquad (2.30)$$

It follows that

$$\Delta U = \Delta H - RT\Delta n_{gas} = \Delta H - 2RT = 176460 - 2 \times 8.31 \times 298 =$$
$$171507 \ J/mol = 171.507 \ kJ/mol.$$

Since reaction proceeds at constant pressure and temperature, heat of reaction is equal:

$$Q_P = \Delta H = 176.460 \ kJ/mol.$$

Problem № 2.11

The reaction proceeds in the autoclave at constant temperature:

$$NH_4Cl_{(solid)} \rightarrow NH_3 + HCl$$

What are the values of ΔH, ΔU and heat of reaction?

Solution:

From the ideal gas equation:

$$\Delta(PV) = R\Delta(Tn)$$

In conditions $T,V = const$

$$V\Delta P = RT\Delta n_{gas}$$

Hence,

$$\Delta H = \Delta U + \Delta(PV) = \Delta U + RT\Delta n_{gas} \qquad (2.30)$$

Following to problem № 2.10, $\Delta H = 176460 \, J/mol$, $\Delta n_{gas} = 2$.
Then
$\Delta U = \Delta H - RT\Delta n_{gas} = 176460 - 2 \times 8.31 \times 298 = 171507 \, J/mol =$
$= 171.507 \, kJ/mol$.
Since reaction proceeds in the autoclave at conditions $V,T = const$, heat of reaction is:

$$Q_V = \Delta U = 171.507 \, kJ/mol.$$

Remark: It is shown from the comparison of results of problems №№ 2.10 and 2.11, that the values of ΔH and ΔU do not depend on the conditions ($P,T = const$ or $V,T = const$) of proceeding of process. It should be so really, since U and H represent state functions and their values do not depend on the way of process.

As to heat, its value is depended on conditions of proceeding of the process:

$$Q_V = \Delta U = 171.507 \; kJ/mol, \qquad when \; V,T = const$$

and

$$Q_P = \Delta H = 176.460 \; kJ/mol, \qquad when \; P,T = const.$$

Problem № 2.12

The molar heat of evaporation of benzol at 353 K and 273 K are equal to 30800 J / mol and 33733 J / mol respectively.

Determine: 1) the value of molar ΔU at 273 and 353 K; 2) the change of system's volume by evaporation of benzol at 353 K; 3) the difference between average heat capacities of liquid benzol and its vapor in the temperature range $273 \div 353$ K.

Solution:

The evaporation of benzol may be expressed so:

$$C_6H_6 \text{ (liq.)} \rightarrow C_6H_6 \text{ (vap.)} \tag{I}$$

1) Let us use the expression (2.30):

$$\Delta H = \Delta U + RT\Delta n_{gas}$$

Hence,

$$\Delta U = \Delta H - RT\Delta n_{gas}$$

At the evaporation of benzol $\Delta n_{gas} = 1 - 0 = 1$. Then

$$\Delta U = \Delta H - RT$$

a)   273 K

$$\Delta U = 33733 - 8.31 \times 273 = 33733 - 2269 = 31464 \; J / mol.$$

b)   353 K

$$\Delta U = 30800 - 8.31 \times 353 = 30800 - 2933 = 27867 \; J / mol.$$

As is seen, the difference between ΔH and ΔU of evaporation of benzol reaches a sufficiently considerable value $(7 \div 10\%)$ at both temperatures. This is stipulated by significant difference between molar volumes of liquid benzol and its vapor.

2) $\Delta \overline{V}$ of evaporation of benzol at 353 K may be described so:

$$\Delta \overline{V} = \overline{V}_2 - \overline{V}_1 = \overline{V}_{vap.} - \overline{V}_{liq.}$$

Let us assume, that benzol is an ideal gas. The molar volume of its vapor may be determined from ideal gas equation:

$$\overline{V}_{benz.vap.} = \frac{RT}{P} = \frac{0.082 \times 353}{1} = 28.94 \ L / mol$$

Then

$$\Delta \overline{V}_{benz.} = \Delta \overline{V}_{benz.vap.} - \Delta \overline{V}_{benz.liq.} = 28.946 - 0.089 = 28.857 \ L / mol.$$

Accordingly, the volume of system by evaporation of one mole of benzol increases substantially (by ~ 29 L).

3) At the evaporation of benzol in the temperature range 273÷573 K:

$$\Delta \overline{C}_{P\,av.} = \overline{C}_{P\,av.(vap.)} - \overline{C}_{P\,av.(liq.)} = const$$

Then Kirchhoff's law may be expressed so:

$$\Delta H_{T_2} = \Delta H_{T_1} + \Delta \overline{C}_p \Delta T \qquad (2.26)$$

From here on:

$$\Delta H_{353} = \Delta H_{273} + \Delta \overline{C}_{P\,av.} (353 - 273)$$

and

$$\Delta \overline{C}_{P\,av.} = \frac{\Delta H_{353} - \Delta H_{273}}{353 - 273} = \frac{30800 - 33733}{80} = -36.66 \ J / K \times mol.$$

Thus, $\Delta \overline{C}_{P\,av.} < 0$. Therefore heat of the considered endothermic process: $C_6H_6 \, _{(liq.)} \rightarrow C_6H_6 \, _{(vap.)}$ decreases with increasing of temperature.

Problem № 2.13

The combustion of propane proceeds in the metallic reactor at 300^0C. The volume of a reactor is 0.6 L. The pressure of the initial mixture is equal to external pressure (1atm). The components present in the mixture with stoichiometric ratio.

1) How does the difference between ΔH and ΔU equal by carrying out a process in: a) isobaric–isothermal and b) isochoric–isothermal conditions?

2) What is the difference between ΔH and ΔU, if 1 mole of C_3H_8 takes part in the reaction?

Solution:

$$C_3H_8 + 5O_2 \rightarrow 3CO_2 + 4H_2O$$

1a) The difference between ΔH and ΔU in isobaric process according to expression (1.26) is equal:

$$\Delta H - \Delta U = P\Delta V = P(V_{fin.} - V_{init.}) = P(V_{prod.} - V_{react.})$$

where $V_{react.}$ is the volume of mixture of reactants and $V_{prod.}$ represents the volume of mixture of products.

According to condition of problem, $P = 1\ atm$ and $V_{init.} = 0.6\ L$. Let us assume each substance participating in the reaction as ideal gas. Then the number of moles in the initial mixture according ideal gas equation is equal:

$$n_{init.} = \frac{PV_{in}}{RT} = \frac{1 \times 0.6}{0.082 \times 573} = 1.28 \times 10^{-2}\ mole.$$

Let us find the number of moles of products. The relationship between products and reagents, according equation of reaction is:

$$\frac{n_{prod.}}{n_{react}} = \frac{7}{6}$$

From here on $n_{prod.} = \frac{7}{6} n_{react.} = \frac{7}{6} \times 1.28 \times 10^{-2} = 1.49 \times 10^{-2}\ mole.$

For the determination of volume of products' mixture we refer to ideal gas equation:

$$V_{prod.} = \frac{n_{prod.}RT}{P} = \frac{1.49 \times 10^{-2} \times 0.082 \times 573}{1} = 0.7\ L$$

and

$$\Delta H - \Delta U = P\Delta V = 1(0.7 - 0.6) = 0.1\ L \times atm = 0.1 \times 101.34 = 10.134\ J$$

1b) Following equation (1.24) in isochoric process:

$$\Delta H - \Delta U = V\Delta P = V\,(P_{prod.} - P_{react.})$$

where V represents the initial volume of mixture of gases, which is equal to volume of reactor; $P_{react.}$ is a pressure of mixture of reactants and $P_{prod.}$ represents pressure of mixture of products.

Let us use ideal gas equation for the determination of $P_{prod.}$:

$$P_{prod.} = \frac{n_{prod.}RT}{V} = \frac{1.49 \times 10^{-2} \times 0.082 \times 573}{0.6} = 1.17\ atm$$

and

$$\Delta H - \Delta U = V\Delta P = 0.6(1.17 - 1) = 0.6 \times 0.17 = 0.102\ L \times atm =$$

$$= 0.102 \times 101.34 = 10.337\ J.$$

Thus, the difference between ΔH and ΔU in isobaric–isothermal and isochoric–isothermal processes is the same. Actually,

$$\Delta(PV) = \Delta(RTn_{gas}) \tag{1.23}$$

In isobaric–isothermal process:

$$P\Delta V = RT\Delta\,n_{gas}$$

In isochoric–isothermal process:

$$V\Delta P = RT\Delta \, n_{gas}.$$

i.e. in both cases:

$$\Delta H - \Delta U = RT\Delta n_{gas} = 8.31 \times 573 \times 10^{-2}(1.49 - 1.28) = 10 \, J.$$

2) As is already mentioned, in isothermal process (it is insignificant if the process is simultaneously isobaric or isochoric) :

$$\Delta H - \Delta U = RT\Delta \, n_{gas} \qquad (2.30)$$

If one mole of propane takes part in the reaction, in order to establish the value of Δn_{gas}, the equation of reaction may be used. Hence,

$$\Delta n_{gas} = (n_{CO_2} + n_{H_2O}) - (n_{C_3H_8} + n_{O_2}) = (3 + 4) - (1 + 5) = 7 - 6 = 1$$

and

$$\Delta H - \Delta U = RT\Delta \, n_{gas} = 8.31 \times 573 \times 1 = 4762 \, J = 4.762 \, kJ.$$

Problem № 2.14

Does energy of so–called "expanded" system increase or decrease, if the reaction:

$$CaCO_3 \rightarrow CaO + CO_2$$

will be carried out at 25^0C and 1 atm pressure in mentioned system?
What is heat of this reaction, if a change of internal energy is equal to 174.91 kJ / mol?

Solution:

The "expanded" system represents the combination of thermodynamic system and a piston with weights. Its energy is described so:

$$H = U + PV \qquad (1.9)$$

where U is the internal energy of thermodynamic system, PV represents so–called "energy of piston" * (it is considered as potential energy).

Thus, H represents a sum of internal energy of system and potential energy of piston. Its change equals:

$$\Delta H = \Delta U + \Delta(PV) = \Delta U + P\Delta V + V\Delta P \qquad (1.12)$$

According to condition of problem, a reaction proceeds at constant pressure and temperature $P,T = const$. In this case a change of energy of "expanded" system is:

$$\Delta H = \Delta U + P\Delta V \qquad (1.29)$$

The internal energy of thermodynamic system in conditions of $P,T = =const$ may be expressed so:

$$\Delta U = Q_P - P\Delta V \qquad (a)$$

Let us assume, that CO_2 is an ideal gas in given conditions, then

$$\Delta(PV) = \Delta(RTn_{gas}) \qquad (1.23)$$

From here on it is obtained for isothermal–isobaric conditions:

$$P\Delta V = RT\Delta n_{gas}$$

A change of number of moles in gaseous phase according to equation of reaction equals:

$$\Delta n_{gas} = 1 - 0 = 1 > 0.$$

Thus, as a result of reaction the system must expand. Work of expansion is equal:

$$P\Delta V = RT\Delta n_{gas} = 8.31 \times 298 \times 1 = 2476.380 \ J = 2.476 \ kJ.$$

---

*In some cases PV is called as volume energy of system.

Internal energy of thermodynamic system reduces by 2.467 *kJ* at the expansion {see expression (a)}, but potential energy of piston increases by 2.476 *kJ*. Consequently for "expanded" system:

$$\Delta H = \Delta U + P\Delta V = 174.910 + 2.480 = 177.386 \ kJ.$$

Heat of isobaric process according to expression (a) is:

$$Q_P = \Delta U + P\Delta V = 174.4910 + 2.476 = 177.386 \ kJ.$$

Thus, change of enthalpy ΔH represents change of energy of "expanded" system on the one hand, but on the other hand it is a heat (Q_P) of isobaric process, proceeding in thermodynamic system.

Problem № 2.15

The reaction

$$2H_2 + CO \rightarrow CH_3OH_{(liq.)}$$

proceeds in a cylinder with mobile piston at 25^0C and 1 atm pressure. The reaction is exothermic and its heat in the conditions of constant volume is: $Q_V = -120 \ kJ/mol$.

1) Does the system expand or compress as a result of reaction?

2) Does a potential energy of piston increase or decrease?

3) Which quantity will reduce more: internal energy or enhalpy of system?

Solution:

1)   The number of moles in a gaseous phase reduces at proceeding of the process:

$$\Delta n_{gas} = 0 - 3 = -3 < 0.$$

According to condition of problem, at proceeding of reaction $P,T = =const$, but system may change a volume.

At $P,T = const$ from ideal gas equation:

$$P\Delta V = RT\Delta n_{gas} = 8.31 \times 298 \times (-3) = -7429.140 \ J = -7.429 \ kJ.$$

$P\Delta V < 0$ means, that $V_2 < V_1$ or system compresses at proceeding of process.

2) Compression is accomplished by piston. Work, expended on the compression is:

$$A_{compr.} = P\Delta V = 7.429 \ kJ.$$

Due to this, energy of piston decreases by 7.429 *kJ*. Actually, the height of piston's site reduces as a result of compression, which stipulates a decrease of potential energy of piston.

3) Since reaction is exothermic, both internal energy and enthalpy reduce at its proceeding. But enthalpy reduces more significant, because its change is conditioned by reducing of both internal energy of system and potential energy of piston.

At $P,T = const$, according expression (1.29):

$$\Delta H = \Delta U + P\Delta V = \Delta U + RT\Delta n_{gas}$$

Following condition of problem, $Q_V = \Delta U = -120 \ kJ/mol$. Then

$$\Delta H = \Delta U + RT\Delta n_{gas} = -120000 + 8.31 \times 298 \times (-3) = -127429 \ J/mol$$
$$= 127.429 \ kJ/mol.$$

Thus, as a result of reaction, enthalpy reduces significantly more, than internal energy.

Problem № 2.16

The patient with a raised temperature (also a healthy person in the hot summer days) suffers from running with sweat. The organism protects oneself from overheating by this way.

What amount of liquid must excrete a patient as sweat in order to decrease the temperature of its body from 40°C to 36.5°C? Assume, that heat capacity of body is equal to 4.18 J / K×g, weight of body is 80 kg and molar heat of evaporation of water at 298 K is equal to 44 kJ / mol.

Solution:

The heat capacity of body with weight 80 *kg* is equal:

$$C_{body} = 80 \times 10^3 \times 4.18 = 334.4 \times 10^3 \ J/K = 334.4 \ kJ/K.$$

Energy, expended by body on the evaporation of water is:

$$q = C_{body} \times \Delta T_{body} = 334.4(40 - 36.5) = 1170.4 \ kJ.$$

On the other hand,

$$q = n\overline{\Delta H}_{evapor.}$$

where n is the number of moles excreted by organism and $\overline{\Delta H}_{evapor.}$ represents molar heat of evaporation of water.
Hence,

$$n_{H_2O} = \frac{q}{\overline{\Delta H}_{evapor.}} = \frac{1170.4}{44} = 26.6 \ mol$$

$$m_{H_2O} = 26.6 \times 18 = 478.8 \ g$$

Problem № 2.17
The solution contains 1 gram equivalent of potassium hydroxide. This solution was added to the very diluted solution, which includes 1 gram equivalent of hydrogen chloride and 1 gram equivalent of acetic acide. Due to neutralization, potassium chloride and potassium acetate were obtained with ratio 3 : 1.
What amount of heat releases as a result of process described above, if heat of dissociation of acetic acid is equal to –0.17 kJ / mol?

Solution:
Heat of neutralization of strong acids with strong alkalis is the same and equals to –55.9 *kJ / mol*. Heat of neutralization of weak acids besides this value (–55.9 *kJ / mol*) contains also heat of its dissociation. Thus, heat of neutralization of acetic acid is:

$$\Delta H_{neutr.CH_3COOH} = -55.9 + \Delta H_{diss.CH_3COOH} = -55.9 - 0.17 = -56.07 \ kJ/mol$$

According to condition of problem, neutralization proceeds with 1 gram equivalent potassium hydroxide. In this case products are obtained in following ratio:

$$KCl : CH_3COOK = 3 : 1$$

This means, that 0.75 gram equivalent of HCl and 0.25 gram equivalent of CH_3COOH enter the neutralization reaction. The amount of heat released at neutralization is:

$$0.75 \, (-55.9) + 0.25 \, (-56.07) = -41.925 - 14.018 = -55.943 \, kJ$$

Problem № 2.18

The values of heat of dissolution for magnesium sulphate's crystallohydrates are given:

(1) $MgSO_4 \cdot H_2O + aq \rightarrow MgSO_4 \cdot aq$ $\Delta H_1 = -55.64 \, kJ / mol$

(2) $MgSO_4 \cdot 7H_2O + aq \rightarrow MgSO_4 \cdot aq$ $\Delta H_2 = 15.90 \, kJ / mol$

Determine the value of heat of reaction (3):

$$MgSO_4 \cdot H_2O + aq \rightarrow MgSO_4 \cdot 7H_2O + aq$$

Solution:

Reactions (1) and (2) are characterized with the same final and different initial states. Therefore heat of intertransformation of their initial states {or heat of reaction (3)} is equal to the difference between heats of processes (1) and (2).

Actually, let us substract reaction (2) from reaction (1):

   (1) $MgSO_4 \cdot H_2O + aq \rightarrow MgSO_4 \cdot aq$

–

   (2) $MgSO_4 \cdot 7H_2O + aq \rightarrow MgSO_4 \cdot aq$

_____

   (3) $MgSO_4 \cdot H_2O + aq \rightarrow MgSO_4 \cdot 7H_2) + aq$

Following Hess's law,

$$\Delta H_3 = \Delta H_1 - \Delta H_2 = -55.64 - 15.90 = -71.54 \; kJ / mol.$$

Problem № 2.19

50 g of water was volatilized from 500 g of crystallohydrate $CuSO_4 \cdot 5H_2O$. Heat of dissolution of anhydrous copper sulphate $(CuSO_4)$ is equal to -66.462 kJ / mol, but heat of dissolution of crystallohydrate $(CuSO_4 \cdot H_2O)$ is 11.704 kJ / mol.

    1) What is the heat of dissolution of residual salt?

    2) What is the percentage of water in the weathered salt?

Solution:

    1) Let us determine the initial amount of water in crystallohydrate:

$$CuSO_4 \cdot 5H_2O \text{————} 5H_2O$$

$$250 \; g \; CuSO_4 \cdot 5H_2O \text{————} 90 \; g \; H_2O$$

$$500 \; g \quad `` \quad `` \text{——} x \; g \; ``$$

$$x = \frac{500}{250} \times 90 = 180 \; g \; H_2O$$

The amount of water remained after the partial weathering of crystallohydrate is:

$$180 - 50 = 130 \; g \; H_2O$$

To this amount of remained water corresponds:

$$\frac{250}{90} \times 130 = 361.1 \; g \; \text{ or } \; \frac{361.1}{250} = 1.44 \; mol \; CuSO_4 \cdot 5H_2O.$$

The amount of anhydrous copper sulphate $(CuSO_4)$, which corresponds to 50 g volatilized water is:

$$\frac{160}{90} \times 50 = 88.8 \ g \quad or \quad \frac{88.8}{160} = 0.55 \ mol \quad CuSO_4.$$

Thus, 0.55 *mole* of $CuSO_4$ and 1.44 *mole* of $CuSO_4 \cdot 5H_2O$ present in the partially weathered salt. Heat of dissolution of this salt equals:

$$0.55 \times (-66.462) + 1.44 \times 11.704 = -19.7 \ kJ \ / \ mol.$$

2) The mass of the remained salt is equal to: $500 - 50 = 450 \ g$. It contains 130 g water (vide supra). The percentage of water in the partially weathered crystallohydrate is:

$$\frac{130}{450} \times 100 = 28.88 \ \%.$$

Problem № 2.20
Use data of problem № 2.19 and calculate ratio between $CuSO_4 \cdot 5H_2O$ and $CuSO_4$, which results in vanishing heat of dissolution of partially weathered crystallohydrate. Express this ratio by using of : a) molar fractions, b) weight per cents.

Solution:
Let us designate the number of moles of $CuSO_4 \cdot 5H_2O$ via n_1, but number of moles of $CuSO_4$ by n_2. Equation of thermal balance of dissolution of partially weathered crystallohydrate in water may be expressed so:

$$n_1 \times 11.704 + n_2 \times (-66.462) = 0$$

$$11.704 \ n_1 - 66.462 \ n_2 = 0$$

$$11.704 \ n_1 = 66.462 \ n_2$$

$$\frac{n_1}{n_2} = \frac{66.462}{11.704} = 5.68$$

Thus, if 5.7 moles $CuSO_4 \cdot 5H_2O$ account for 1 mole $CuSO_4$, heat of dissolution of partially weathered crystallohydrate in water will be equalled to zero.

Let us express this ratio by:

a) molar fractions.

$$\kappa_1 = \frac{n_1}{n_1 + n_2} = \frac{5.7}{5.7 + 1} = 0.85$$

$$\kappa_2 = \frac{n_2}{n_1 + n_2} = \frac{1}{5.7 + 1} = 0.15$$

b) weight per cents.

The mass of 5.7 *moles* $CuSO_4 \cdot 5H_2O$ is equal to $5.7 \times 250 = 1425$ g. The mass of 1 *mole* $CuSO_4$ is 160 g. net mass is: $1425 + 160 = 1585$ g.

$$\%CuSO_4 \cdot 5H_2O = \frac{1425}{1585} \times 100 = 89.9 \%$$

$$\%CuSO_4 = \frac{160}{1585} \times 100 = 10.1 \%$$

Problem № 2.21

Heat of dissolution of gaseous and liquid ammonia at 298 K are −34.903 kJ / / mol and −11.22 kJ / mol respectively. What is the heat of formation of liquid ammonia, if the heat of formation of gaseous ammonia equals to −46.19 kJ / mol?

Solution:

Let us express heat of formation of liquid ammonia from gaseous one by using of heats of their dissolution:

1) $NH_{3\,(g)} + aq \rightarrow NH_3 \cdot aq$ $\Delta H_1 = -34.903 \; kJ/mol$

$-$

2) $NH_{3\,(liq.)} + aq \rightarrow NH_3 \cdot aq$ $\Delta H_2 = -11.22 \; kJ/mol$

_____

3) $NH_{3\,(g)} \rightarrow NH_{3\,(liq.)}$

Following Hess's law:

$$\Delta H_3 = \Delta H_1 - \Delta H_2 = -34.903 - (-11.22) = -23.683 \ kJ \, / \, mol.$$

The formatin of liquid ammonia may be represented so:

4) $\frac{1}{2} N_2 + 3/2 \, H_2 \rightarrow NH_{3 \, (g)}$ $\qquad\qquad\qquad \Delta H_4 = -46.19 \ kJ \, / \, mol$

+

3) $NH_{3 \, (g)} \rightarrow NH_{3 \, (liq.)}$ $\qquad\qquad\qquad \Delta H_3 = -23.68 \ kJ \, / \, mol$

───────────────────────

5) $\frac{1}{2} N_2 + 3/2 \, H_2 \rightarrow NH_{3 \, (liq.)}$

$$\Delta H_5 = \Delta H_{form \, .liq.ammon.} = \Delta H_4 + \Delta H_3 = -46.19 - 23.68 = -69.87 \ kJ \, / \, mol.$$

Problem № 2.22
Determine heat of dilution of hydrochloric acid with concentration 33.64% to 1%.

Solution:
The values of integral heat of dissolution are given in a handbook. The concentrations are expressed with number of moles of water per one mole of HCl. Therefore concentrations must be taken from percentage to molar.

100 *g* solution of 33.64% HCl contains 33.64 *g* of HCl and $100 - 33.64$ = 66.36 *g* of H_2O. This makes up $\dfrac{33.64}{36.5} = 0.92$ *mole* of HCl and $\dfrac{66.36}{18} =$ = 3.69 *mole* of H_2O.

0.92 *mole* of HCl account for 3.69 *mole* of H_2O

1 *mole* of HCl " " *x* *mole* of H_2O

$$x = \frac{3.69}{0.92} = 4 \ moles \ of \ H_2O$$

b) 100 g solution of 1% HCl contains 1 g of HCl and 99 g of H_2O, which correspond to $\dfrac{1}{36.5} = 0.027$ *mole* of HCl and $\dfrac{99}{18} = 5.5$ *mole* of H_2O.

0.027 *moles* of HCl account for 5.5 *moles* of H_2O

1 *mole* of HCl account for x *moles* of H_2O

$$x = \frac{5.5}{0.027} = 203.7 \approx 200 \; moles \; of \; H_2O$$

Thus, in the solution of 1% HCl one *mole* of HCl accounts for 200 *mole* of water.

The integral heats of dissolution are found in the handbook:

% HCl	Number of moles of water per one mole of HCl	$\Delta H_{dissolution}$ kJ / mol
33.64	4	–61.2
1	~200	–74.2

Hence

$$\Delta H_{dilution} = -74.2 - (-61.2) = -13 \; kJ \, / \, mol.$$

Problem № 2.23

Determine the value of molar isobaric heat capacity of oxygen at 298 and 500 K.

Solution:

Let us use the series of dependence of heat capacity on temperature for oxygen:

$$C_{P(O_2)} = 26.19 + 11.49 \times 10^{-3} \, T - 3.22 \times 10^{-6} \, T^2$$

a) 298 K
$C_{P(O_2)} = 26.19 + 11.49 \times 10^{-3} \times 298 - 3.22 \times 10^{-6} \times 298^2 =$
$= 26.19 + 3.42 - 0.29 = 29.32 \ J/K \times mol.$

b) 500 K
$C_{P(O_2)} = 26.19 + 11.49 \times 10^{-3} \times 500 - 3.22 \times 10^{-6} \times 500^2 =$
$= 26.19 + 5.75 - 0.81 = 31.13 \ J/K \times mol$

Problem № 2.24
What is the value of mean heat capacity of oxygen in the range of
298÷500 K ?

Solution:
The expression is used for determination of average heat capacity:

$$\overline{c}_{p \, av.} = \frac{1}{T_2 - T_1} \int_{T_1}^{T_2} \overline{c}_p \, dT \qquad (2.24)$$

In the handbook we found:

$$C_{P(O_2)} = 26.19 + 11.49 \times 10^{-3} \, T - 3.22 \times 10^{-6} \, T^2$$

Let us introduce this expression into (2.24):

$$\overline{c}_{p \, av.} = \frac{1}{T_2 - T_1} \int_{T_1}^{T_2} \overline{c}_p \, dT = \frac{1}{500 - 298} \times$$

$$\left[26.19(500 - 298) + \frac{11.49 \times 10^{-3}}{2} (500^2 - 298^2) - \frac{3.22 \times 10^{-6}}{3} (500^3 - 298^3) \right] =$$

$$= \frac{1}{202} (5290.38 + 926.07 - 105.76) = 30.25 \ J/K \times mol$$

Problem № 2.25

Determine the change of enthalpy at heating of 100 g of methanol from 500 to 800 K under 1 atm pressure.

Solution:

The series of dependence of molar heat capacity of gaseous methanol on temperature is found in the handbook:

$$C_{P(CH_3OH)_{gas}} = 15.28 + 105.2 \times 10^{-3}\, T - 31.04 \times 10^{-6}\, T^2.$$

The change of enthalpy at heating of 1 mole substance is expressed so:

$$H_{T_2} - H_{T_1} = \int_{T_1}^{T_2} C_P dT$$

$$H_{800} - H_{500} = \int_{500}^{800} (15.28 + 105.2 \times 10^{-3}\, T - 31.04 \times 10^{-6}\, T^2)\, dT =$$

$$= 15.28(800 - 500) + \frac{105.2 \times 10^{-3}}{2}(800^2 - 500^2) - \frac{31.04 \times 10^{-6}}{3} \times$$

$$\times (800^3 - 500^3) = 4584 + 20514 - 4004 = 21094\ J = 21.094\ kJ.$$

100 g of methanol contains $\dfrac{100}{32} = 3.125$ *moles*. Thus, the change of enthalpy of 100 g methanol at heating from 500 to 800 K is:

$$3.125 \times 21.094 = 65.919\ kJ.$$

Problem № 2.26

Heat of reaction:

$$CO + H_2O_{(gas)} \rightarrow CO_2 + H_2$$

at 500 K is equal to −39.794 kJ.

Determine the value of ΔU at 298 K.

Solution:

Let us determine the value of ΔH of reaction at 298 K. For this purpose Kirchhoff's equation is used:

$$\Delta H_{T_2} = \Delta H_{T_1} + \int_{T_1}^{T_2} \Delta C_p dT \qquad (2.21)$$

For the given reaction $\Delta C_P = C_{P(CO_2)} + C_{P(H_2)} - C_{P(CO)} - C_{P(H_2O) \, gas}$.
In the handbook may be found:

$C_{P(CO)} = 28.41 + 4.1 \times 10^{-3} T - 0.46 \times 10^5 T^{-2}$

$C_{P(H_2O)_{gas}} = 30 + 10.71 \times 10^{-3} T + 0.33 \times 10^5 T^{-2}$

$C_{P(CO_2)} = 44.14 + 9.04 \times 10^{-3} T - 8.53 \; 10^5 T^{-2}$

$C_{P(H_2)} = 27.28 + 3.26 \times 10^{-3} T + 0.5 \times 10^5 T^{-2}$

Hence

$\Delta a = 27.28 + 44.14 - 30 - 28.41 = 13.01$

$\Delta b = (3.26 + 9.04 - 10.71 - 4.10) \times 10^{-3} = -2.51 \times 10^{-3}$

$\Delta c' = (0.50 - 8.53 - 0.33 + 0.46) \times 10^5 = -7.90 \times 10^5$

Thus, for reaction:

$$CO + H_2O_{(gas)} \rightarrow CO_2 + H_2$$

$$\Delta C_P = 13.01 - 2.51 \times 10^{-3} T - 7.9 \times 10^5 T^{-2}$$

Let us introduce this expression into (2.21):

$$\Delta H_{298} = \Delta H_{500} + \int\limits_{500}^{298} (13.01 - 2.51 \times 10^{-3}\, T - 7.9 \times 10^{5}\, T^{-2})\, dT =$$

$$= -39794 + 13.01(298 - 500) - \frac{2.51 \times 10^{-3}}{2}\,(298^2 - 500^2) +$$

$$+ 7.9 \times 10^{5} \times \left(\frac{1}{298} - \frac{1}{500} \right) = -39764 - 2628 + 202 + 1071 =$$

$$= -41119\ J/mol = -41.119\ kJ./mol$$

For determination the value of ΔU_{298} we use expression (2.30):

$$\Delta H = \Delta U + RT\Delta\, n_{gas}.$$

In the considered reaction

$$\Delta n_{gas} = 1 + 1 - 1 - 1 = 0.$$

So

$$\Delta H_{298} = \Delta U_{298} = -41.119\ kJ/mol$$

Accordingly, the value of heat of exothermic reaction :

$$CO + H_2O_{gas} \rightarrow CO_2 + H_2$$

reduces with increasing of temperature, which is stipulated by $\Delta C_P > 0$ in the temperature range $298 \div 500$ K.

Problem № 2.27

The standard heats of formation of carbon monoxide and phosgene at 298 K are −110. kJ / mol and −223 kJ / mol.

Determine ΔH at 700 K for reaction:

$$CO + Cl_2 \rightarrow COCl_2$$

Let us determine the value of ΔH at 298 K. According to one of results of Hess's law:

$$\Delta H_{reaction} = \Delta H_{form.COCl_2} - \Delta H_{form. CO} = -223 - (-110.5) = -112.5 \; kJ / mol$$

Let us calculate the value of ΔH of process at 700 K. In the given reaction:

$$\Delta C_P = C_{P(COCl_2)} - C_{P(CO)} - C_{P(Cl_2)}.$$

In the handbook is found:

$$C_{P(CO)} = 28.41 + 4.1 \times 10^{-3} \, T - 0.46 \times 10^5 \, T^2$$

$$C_{P(Cl_2)} = 36.69 + 1.05 \times 10^{-3} \, T - 2.52 \times 10^5 \, T^{-2}$$

$$C_{P(COCl_2)} = 67.16 + 12.11 \times 10^{-3} \, T - 9.03 \times 10^5 \, T^{-2}$$

$$\Delta a = 67.16 - 36.69 - 28.41 = 2.06$$

$$\Delta b = (12.11 - 1.05 - 4.1) \times 10^{-3} = 6.96 \times 10^{-3}$$

$$\Delta c' = (-9.03 + 2.52 + 0.46) \times 10^5 = - 6.05 \times 10^5$$

It follows, that ΔC_P of reaction is:

$$\Delta C_P = 2.06 + 6.96 \times 10^{-3} \, T - 6.05 \times 10^5 \, T^{-2}.$$

Let us introduce obtained result into Kirchhoff's equation:

$$\Delta H_{T_2} = \Delta H_{T_1} + \int\limits_{T_1}^{T_2} \Delta C_P \, dT \qquad (2.21)$$

$$\Delta H_{700} = \Delta H_{298} + \int\limits_{298}^{700} (2.06 + 6.96 \times 10^{-3} \, T - 6.05 \times 10^5 \, T^{-2}) \, dT =$$

$$= -112500 + 2.06 \,(700 - 298) + \frac{6.96 \times 10^{-3}}{2} \,(700^2 - 298^2) +$$

$$+ 6.05 \times 10^5 \left(\frac{1}{700} - \frac{1}{298} \right) = -112500 + 828.12 + 1396.16 - 1165.92 =$$

$$= -111441.63 \, J / mol.$$

Problem № 2.28

The specific heat capacities of graphite and diamond in the temperature range 273÷298 K are equal to 720.83 J / kg×K and 505.58 J / kg×K respectively. Heat of transformation of graphite into diamond at 298 K is equal to 1.9 kJ / mol.

How does the value of heat of transformation: *graphite → diamond* change at decreasing of temperature from 298 K to 273 K?

Solution:

According to Kirchhoff's law:

$$\Delta H_{T_2} = \Delta H_{T_1} + \int\limits_{T_1}^{T_2} \Delta C_P \, dT \qquad (2.21)$$

Since the temperature range is narrow, it may be assumed, that $\Delta C_P = $ =*const*. Then

$$\int_{T_1}^{T_2} \Delta\, C_P dT = \Delta C_p\, \Delta T$$

and

$$\Delta H_{T_2} = \Delta H_{T_1} + \Delta C_P\, \Delta T \tag{2.26}$$

In the considering reaction: *graphite* $\rightarrow$ *diamond*

$$\Delta C_P = C_{P\,(diamond)} - C_{P\,(graphite)} = 505.58 - 720.83 = -215.25\ J/kg \times K$$

Let us transform the difference of specific heat capacities into the molar one:

$$1\ kg\ C = \frac{1000}{12} = 83.33\ mole\ \text{of carbon}$$

$$\Delta C_p = C_{P\,diam.} - C_{P\,graph.} = -\frac{215.25}{83.33} = -2.58\ J/mol \times K.$$

Let us introduce this value of ΔC_p into expression (2.26):

$$\Delta H_{273} = \Delta H_{298} - 2.58\,(273 - 298) = 1900 + 64.5 = 1964.5\ J/mol =$$

$$= 1.964\ kJ/mol.$$

Thus, the value of heat of considered endothermic reaction decreases with increasung of temperature. This is conditioned by $\Delta C_{P\,reaction} < 0$.

Problem № 2.29
Heat of reaction:

$$C + 1/2\ O_2 \rightarrow CO$$

at 298 K is equal to -110.5 kJ / mol. The expression of ΔC_P for this reaction is following:

$$\Delta C_P = 9.91 - 20.78 \times 10^{-3} T + 6.63 \times 10^{-6} T^2.$$

How does the value of heat change in the temperature range 298÷700 K (at 400, 500, 550, 587, 600 and 700 K)? Tabulate the obtained data.

Solution:

Let us divide the considering temperature range into comparatively small intervals, where the use of Kirchhoff's approximate equation is possible:

$$\Delta H_{T_2} = \Delta H_{T_1} + \Delta C_P \, \Delta T \qquad (2.26)$$

1) 298÷400 K

$\Delta C_{P\,298} = 9.91 - 20.78 \times 10^{-3} \times 298 + 6.63 \times 10^{-6} \times 298^2 = 9.91 - 6.19 +$

$+ 0.59 = 4.31 \; J / K\times mol.$

$\Delta H_{400\,K} = \Delta H_{298\,K} + 4.31 \, (400 - 298) = -110500 + 439.25 =$

$= -110060.75 \; J / mol.$

2) 400÷500 K

$\Delta C_{P\,400} = 9.91 - 20.78 \times 10^{-3} \times 400 + 6.63 \times 10^{-6} \times 400^2 = 9.91 - 8.31 +$

$+1.07 = 2.66 \; J / K\times mol.$

$\Delta H_{500\,K} = \Delta H_{400\,K} + 2.66 \, (500 - 400) = -110061 + 266 = -109795 \; J / mol$

3) 500÷550 K

$\Delta C_{P\,500\,K} = 9.91 - 20.78 \times 10^{-3} \times 500 + 6.63 \times 10^{-6} \times 500^2 = 9.91 -$

$10.39 + 1.66 = 1.18 \; J / K\times mol.$

$\Delta H_{550\,K} = \Delta H_{500\,K} + 1.18 \, (550 - 500) = -109795 + 58.88 =$

$= - 109736.12 \, J / mol.$

4) $550 \div 587$ K

$\Delta C_{P\,550\,K} = 9.91 - 20.78 \times 10^{-3} \times 550 + 6.63 \times 10^{-6} \times 550^{\,2} = 9.91 -$

$11.43 + 2 = 0.49 \, J / K \times mol.$

$\Delta H_{587\,K} = \Delta H_{550\,K} + 0.49\,(587 - 550) = - 109736 + 18.13 =$

$= - 109717.87 \, J / mol.$

5) $587 \div 600$ K

$\Delta C_{P\,587\,K} = 9.91 - 20.78 \times 10^{-3} \times 587 + 6.63 \times 10^{-6} \times 587^{\,2} = 9.91 -$

$12.2 = 2.28 = - 0.003 \, J / K \times mol.$

$\Delta H_{600\,K} = \Delta H_{587\,K} - 0.003\,(600 - 587) = - 109718 - 0.04 =$

$= - 109718.04 \, J / mol.$

6) $600 \div 700$ K

$\Delta C_{P\,600\,K} = 9.91 - 20.78 \times 10^{-3} \times 600 + 6.63 \times 10^{-6} \times 600^{\,2} = 9.91 -$

$12.47 + 2.39 = - 0.17 \, J / K \times mol.$

$\Delta H_{700\,K} = \Delta H_{600\,K} - 0.17\,(700 - 600) = - 109718.04 - 17 =$

$= - 109735.04 \, J / mol.$

$\Delta C_{P\,700\,K} = 9.91 - 20.78 \times 10^{-3} \times 700 + 6.63 \times 10^{-6} \times 700^{\,2} = 9.91 -$

$-14.55 = 3.25 = - 1.39 \, J / K \times mol.$

Let us tabulate the obtained data.

Table 2.1. Dependence of heat and ΔC_P on the temperature for the reaction:

$$C + \tfrac{1}{2} O_2 \rightarrow CO$$

T K	ΔH, kJ/mol	ΔC_P, J/K×mol
298	– 110.500	4.31
400	– 110.061	2.66
500	– 109.795	1.18
550	– 109.736	0.49
587	– 109.718	– 0.003 ≈ 0
600	– 109.718	– 0.17
700	– 109.735	– 1.39

As is seen from Table 2.1, the absolute value of exothermic reaction : $C + 1/2O_2 \rightarrow CO$ decreases in the temperature range 298÷587 K and increases within 600÷700 K. In the interval of 587÷600 K the value of ΔH is independent of temperature. Such a change of heat with temperature is stipulated by temperature dependence of ΔC_P of reaction: $\Delta C_P \approx 0$ at 587 K, $\Delta C_P > 0$ below this temperature, but $\Delta C_P < 0$ above it.

Problem № 2.30

As a result of combustion of 1 g benzoic acid at 298 K, the temperature of adiabatic flame calorimeter (operates under constant pressure) is risen by 17.76^0. Heat capacity of calorimeter is equal to 1.5 kJ / K

What is the heat of combustion of benzoic acid in the conditions of: a) constant pressure; b) constant volume.

Solution:

$$C_7H_6O_2 + 7.5\ O_2 \rightarrow 7CO_2 + 3\ H_2O_{\ (liq.)}$$

a) Expression (2.11) is used for determination of heat of process in the conditions of constant pressure:

$$q_p = -C_{cal.} \, \Delta T_{cal.} = -1500 \times 17.76 = -26640 \ J.$$

Heat of combustion of benzoic acid is equal:

$$Q_P = 122 \, (-26640) = -3250080 \ J / mol = -3250 \ kJ / mol.$$

b) Let us determine heat of reaction in the conditions of constant volume from expression:

$$Q_V = Q_P - RT\Delta \, n_{gas} \tag{2.30$'$}$$

At the combustion of benzoic acid

$$\Delta n_{gas} = 7 - 7.5 = -0.5$$

$$Q_V = -3250000 - 8.31 \times 298 \times (-0.5) = -3248844.2 \ J / mol =$$

$$= -3249 \ kJ / mol.$$

Problem № 2.31

0.5 g benzoic acid is burnt at 313 K in the same calorimeter (see problem № 2.30).

1) What is the heat of combustion of benzoic acid at 313 K and under constant pressure, if molar $\Delta U = -3246893$ J / mol at the same temperature?

2) How does temperature of calorimeter change at proceeding of the reaction?

Solution:

1) As is known,

$$\Delta H = \Delta U + RT\Delta n_{gas} \tag{2.30}$$

In the given reaction $\Delta n_{gas} = -0.5$ (see problem № 2.30). Thus,

$$\Delta \overline{H} = -3246893 + 8.31 \times 313(-0.5) = -3248193.5 \; J/mol.$$

2) The amount of heat, released in the calorimeter at combustion of 0.5 g benzoic acid is:

$$q_p = \frac{0.5}{122} \times \Delta \overline{H} = \frac{0.5}{122}(-3248194) = -13312.27 \; J$$

On the other hand,

$$q_p = -C_{cal.} \; \Delta T_{cal.}.$$

From here on

$$\Delta T_{cal.} = -\frac{q_p}{-C_{cal.}} = \frac{-13312.27}{-1500} = 8.87 \; K$$

Problem № 2.32

Use the date of problems №№ 2.30 and 2.31 and determine the value of C_P of benzoic acid in the range of temperature 298÷313 K. Assume, that heat capacities of substances, taking part im the reaction of combustion of benzoic acid are invariable in this range of temperature.

Solution:

According to condition of problem, ΔC_p of reaction within 298÷313 K is invariable. Therefore Kirchhoff's law in this range of temperature may be represented so:

$$Q_{pT_2} = Q_{pT_1} + \Delta C_p (T_2 - T_1) \qquad (2.26')$$

Hence,

$$\Delta C_p = \frac{Q_{p313^0} - Q_{p298^0}}{313 - 298} = \frac{-3248195 - (-3250080)}{15} = 125.733 \; J/K \times mol$$

On the other hand, for reaction

$$C_7H_6O_2 \text{ (liq.)} +7.5\ O_2 \rightarrow 3\ H_2O_{\text{(liq.)}} + 7CO_2$$

$$\Delta C_P = 3\ C_{P(H_2O)_{liqs}} + 7\ C_{P(CO_2)} -7.5\ C_{P(O_2)} - C_{P(C_7H_6O_2)_{liq.}}$$

Hence

$$C_{P(C_7H_6O_2)_{liq.}} = 3\ C_{P(H_2O)_{liqs}} + 7\ C_{P(CO_2)} -7.5\ C_{P(O_2)} - \Delta C_P$$

Let us find the values of average heat capacities at 298 K in the hand book:

$$C_{P(H_2O)_{liqs}} = 75.31\ J\ /\ K{\times}mol$$

$$C_{P(CO_2)} = 37.13\ J\ /\ K{\times}mol$$

$$C_{P(O_2)} = 29.36\ J\ /\ K{\times}mol.$$

Then

$$C_{P(C_7H_6O_2)_{liq.}} = 3\ C_{P(H_2O)_{liqs}} + 7\ C_{P(CO_2)} -7.5\ C_{P(O_2)} - \Delta C_P = 3 \times 75.31 +$$

$$+ 7 \times 37.13 - 7.5 \times 29.36 - 125.73 = 225.93 + 259.91 - 220.2 - 125.73 =$$

$$= 139.91\ J\ /\ K \times mol.$$

3. Entropy

Theoretical Part

According to one of the most common law of nature (second law of thermodynamics) the natural processes take place in one direction only. This direction is conditioned by affinity of energy for dissipation i.e. dissipation of energy and particles in the space and transferring of ordered form of motion into disordered one.

The transfer of the heat from the hot body to the cold one proceeds spontaneously due to the energy dissipation. The reversed process – the transfer of the heat from the cold body to the hot body does not take place, because it would induce the concentration of energy on the hot body i.e. opposite phenomenon to the energy dissipation.

The dissipation of particles, carrying energy, takes place during gas expansion in the space. Therefore this process proceeds spontaneously. The reversed process – the compression of gas – is not realized spontaneously, because it would cause the localization of energy in less volume of the space.

The body moved on the horizontal surface is stopped after the definite time. This is conditioned by transferring of the ordered form of the motion into the disordered one. The reversed phenomenon – beginning of the motion of the immobile body – does not take place spontaneously, because it requires the transferring of the disordered (thermal) form of motion into arranged one.

In order to characterize the natural processes, the concept of entropy (S) was proposed and the second law of thermodynamics was formulated by

Clausius (Rudolf Gotlibb) in 19-th century. According to this law, the natural processes take place in direction of entropy increase. The entropy tends to its maximum value, which is reached in the equilibrium state indeed.

It was found that the conceptions of classical thermodynamics were insufficient for the elucidation of entropy essence. By Clausius himself the establishment of the physical concept of entropy would be possible only after the relation of entropy with the properties of the particles constituting the system.

Ludwig Boltzmann, who created the basis of the new field of theoretical physics – statistical thermodynamics, has solved this problem. Boltzmann has correlated entropy with the probability:

$$S = k \ln W \qquad (3.1)$$

where k is Boltzmann's constant ($k = 1.38 \times 10^{-23}$ J/K); W represents thermodynamic probability of macrostate of the system. W is equal numerically to the quantity of microstates, by means of which the given macrostate is realized.

Let us consider the model system consisted of N particles of monoatomic ideal gas. The decisive factor, which stipulates the character of process proceeded in this (or anyone other) system is a tendency of each particle and hence of the system as a whole, to the occupation of all possible energy levels.

Let us assume that in the initial state the thermodynamic probability of (macro)state of the system is W_1; the quantity of microstates W_1 corresponds to it. The system may transfer spontaneously only into the state with higher thermodynamic probability $W_2 > W_1$, which is determined by the larger number of microstates W_2. By Boltzmann, the more energy levels are occupied by particles, the more is the number of microstates, and hence, the value of S of the system. This value is maximum, when all possible energy levels are occupied i.e. at the state of thermodynamic equilibrium. Thus, the increase of entropy occurs in the spontaneous processes: $S \rightarrow S_{max}$ ($S_{max} = S_{equil.}$).

The thermodynamic probability of system's (macro)state is expressed so:

$$W = \frac{N!}{N_1! N_2! ... N_r!} \tag{3.2}$$

where r is the number of energy levels, and N_1, N_2,...N_r are the number of particles, which are populated on the I, II,...r–th energy levels respectively.

The uniform distribution of particles on the energy levels is observed at the equilibrium:

$$W_{max} = \frac{N!}{\left(\left[\frac{N}{r} \right]! \right)^r} \tag{3.3}$$

where r is the number of energy levels.

Stirling's approximate equation is correct for the factorials of large numbers:

$$\ln N! \approx N \ln N - N \tag{3.4}$$

After the equilibrium is established, the system could not transfer spontaneously into the state considerably distant from equilibrium. However, it undergoes so-called small fluctuations (temporal and insignificant deviations from the equilibrium state) from time to time. The reason consists in the chaotic character of particles' motion. The particles change their energy state contin\uneously, which results in permanent transfer of system from one microstate into the other. According to the principle of equal apriori probability of microstates, the system may be found in microstate, which corresponds to the macrostate, distinguished from the equilibrium one i.e. the system undergoes fluctuation. Since the number of microstates, corresponded to the equilibrium ($W_{equi,}$), is incomparably higher than the number or the sum of numbers of other microstates, the system immediately returns to the equilibrium.

Finally it may be concluded that entropy is the value, which characterizes quantitavely the tendency of the matter to the occupation of all possible energy levels i.e. the affinity of energy to dissipation.

In spontaneous processes the devaluation of energy takes place from the standpoint of obtaining work from it. Really, as we have mentioned above, the natural processes are accompanied by the dissipation of energy. Therefore, after all natural processes the energy is transformed into the less available for us form i.e. it is degradated. Thus, the entropy is the measure of both the spontaneity of process and degradation of energy.

The entropy change in the thermodynamic process is expressed so:

$$\Delta S = S_{fin.} - S_{init.} = k \ln \frac{W_{fin.}}{W_{init.}} \qquad (3.5)$$

If the process occurs spontaneously, then $W_{fin.} > W_{init.}$, $S_{fin.} > S_{init.}$ and $\Delta S > 0$.

Let sign:

$$\sigma = \ln W \qquad (3.6)$$

$$\Delta \sigma = \sigma_{fin.} - \sigma_{init.} = \ln \frac{W_{fin.}}{W_{init.}} \qquad (3.7)$$

From the comparison of expressions (3.1) and (3.6), also (3.5) and (3.7) follows:

$$S = k\sigma \qquad (3.8)$$

and

$$\Delta S = k\Delta \sigma \qquad (3.9)$$

The entropy of monoatomic ideal gas is described by Sacure—Tetrode's equations (1912):

$$\sigma = N\left(\ln T^{3/2} - \ln \frac{N}{V} + const \right) \tag{3.10}$$

$$S = kN\left(\ln T^{3/2} - \ln \frac{N}{V} + const \right) \tag{3.11}$$

where N is the number of particles, V represents the volume of system and k is Boltzmann's constant;

$$const = \ln\left(\frac{2\pi mk}{h^2} \right)^{3/2} + \frac{5}{2} \tag{3.12}$$

where m is the mass of particle, and h is Plank's constant ($h = 6.626 \times 10^{-34}$ J × sec).

Taking into account that $kN = Rn$, then from (3.11) expression is obtained:

$$S = Rn\left(\ln T^{3/2} - \ln \frac{N}{V} + const \right) \tag{3.13}$$

When $N = const$, according to (3.10), (3.11), (3.13), the entropy change is expressed by following:

$$\Delta\sigma = N\left(\frac{3}{2} \ln \frac{T_2}{T_1} + \ln \frac{V_2}{V_1} \right) \tag{3.14}$$

$$\Delta S = kN\left(\frac{3}{2} \ln \frac{T_2}{T_1} + \ln \frac{V_2}{V_1} \right) \tag{3.15}$$

$$\Delta S = Rn \left(\frac{3}{2} \ln \frac{T_2}{T_1} + \ln \frac{V_2}{V_1} \right) \qquad (3.16)$$

The introduction of entropy concept and the establishment of the second law of thermodynamics are historically related with heat engines. According to Carnot's theorem, the efficiency of the ideal heat engine is determined by the temperature of heater and refrigerator and not is depend on the nature of working body:

$$\eta = \frac{T_h - T_c}{T_h} = \frac{A}{q_h} \qquad (3.17)$$

where T_h is the temperature of heater, T_c represents the temperature of refrigerator, A is fulfilled work and q_h is heat, taken from the heater.

Carnot's cycle represents the principal thermodynamic cycle, efficiency of which is higher than efficiency of any other cycle, working in the same temperature range. Therefore, in order to establish the upper limit of the efficiency of heat engines, Carnot's cycle is used; however, no real engine is worked by this cycle.

Carrying out Carnot's cycle in the reversed direction, the refrigerative cycle is obtained instead of heat cycle. In the Carnot's reversed cycle the heat transfers from cold body (the refrigerator) to the hot body (the heater). This process is unnatural and the fulfillment of a certain work is required for its realization. Besides, the heater receives not only the heat of refrigerator, but also the energy, equivalent of work A, expended for obtaining this heat:

$$q_h = q_c + A \qquad (3.18)$$

The efficiency of Carnot's reversed cycle (so-called coefficient of cooling) is expressed so:

$$\eta' = \frac{T_c}{T_h - T_c} = \frac{q_c}{A} \qquad (3.19)$$

where q_c is heat taken from refrigerator and A represents work expended for taking this heat.

With further extension of Carnot's theorem Clausius introduced the entropy concept in thermodynamics. The entropy is the state function of system and its change expresses so:

$$dS \geq \frac{\delta q}{T} \tag{3.20}$$

In the isothermal reversible process:

$$\Delta S = \frac{q_{rev.}}{T} \tag{3.21}$$

In the irreversible process:

$$\Delta S > \frac{q_{irrev.}}{T} \tag{3.22}$$

Since entropy is a state function, its changes will be equal in both reversible and irreversible processes (when initial and final states of the system are the same):

$$\Delta S_{rev.} = \Delta S_{irrev.}$$

Entropy represents the additive quantity also. It means that:

a) entropy of the system is equal to the sum of entropies of its constituent parts;

b) entropy of so-called "united" ("combined") system is equal to the sum of entropies of its constituted systems;

c) the entropy change of the system is equal to the sum of entropy changes of its constituents;

d) the entropy change in some process is equal to the sum of entropy changes of the stages, in which given process may be divided.

The change of entropy in reversible isochoric ($V=const$) process is:

$$\Delta S = \int_1^2 dS = \int_1^2 \frac{\delta q_v}{T} = \int_1^2 \frac{dU}{T} \qquad (3.23)$$

In reversible isobaric ($P=const$) process:

$$\Delta S = \int_1^2 dS = \int_1^2 \frac{\delta q_p}{T} = \int_1^2 \frac{dH}{T} \qquad (3.24)$$

The phase transformation proceeds at constant temperature and pressure. Therefore, according to (3.24):

$$\Delta S_{ph.tr.} = \int_1^2 \frac{dH}{T} = \frac{\Delta H_{ph.tr.}}{T_{ph.tr.}} \qquad (3.25)$$

At the heating (cooling) in isochoric conditions from (3.23) is followed:

$$\Delta S_{heat.} = \int_{T_1}^{T_2} \frac{\delta q_v}{T} = n \int_{T_1}^{T_2} \frac{\overline{c}_v}{T} dT \qquad (3.26)$$

If in the given temperature range $\overline{c}_v = const$ then from (3.26) is obtained:

$$\Delta S_{heat.} = n\overline{c}_v \ln \frac{T_2}{T_1} \qquad (3.27)$$

The entropy change of the heating (cooling) in isobaric conditions is equal to:

$$\Delta S_{heat.} = \int_{T_1}^{T_2} \frac{\delta q_p}{T} = n \int_{T_1}^{T_2} \overline{c}_p dT \qquad (3.28)$$

If $\bar{c}_p = const$, then

$$\Delta S_{heat.} = n\bar{c}_p \ln \frac{T_2}{T_1} \tag{3.29}$$

Generally, in the processes with participation of the ideal gas:

$$\Delta S = n \int_{T_1}^{T_2} \bar{c}_p \frac{dT}{T} + nR \ln \frac{V_2}{V_1} \tag{3.30}$$

from which in the cases of $\bar{c}_v = const$ or $\bar{c}_p = const$ we obtain:

$$\Delta S = n\bar{c}_v \ln \frac{T_2}{T_1} + nR \ln \frac{V_2}{V_1} \tag{3.31}$$

and

$$\Delta S = n\bar{c}_p \ln \frac{T_2}{T_1} + nR \ln \frac{P_1}{P_2} \tag{3.32}$$

The expression of entropy change at isothermal expansion (compression) of gas may be deduced from (3.31) and (3.32):

$$\Delta S = nR \ln \frac{V_2}{V_1} = nR \ln \frac{P_1}{P_2} \tag{3.33}$$

As regards to adiabatic processes, they are isentropic. Really, the heat exchange with surroundings is excluded in adiabatic conditions: $q=0$. Therefore in reversible adiabatic process:

$$\Delta S_{ad.} = \int_1^2 dS_{ad.} = \int_1^2 \frac{\delta q_{ad.}}{T} = 0 \tag{3.34}$$

The entropy change at the mixing of ideal gases is expressed by following:

$$\Delta S_{mix.} = -R\sum n_i \ln \kappa_i \qquad (3.35)$$

where n_i is the number of moles of $i-th$ component and κ_i represents the mole fraction of $i-th$ component.

When $\sum n_i = 1$, then $n_i = \kappa_i$, and from expression (3.35) may be obtained:

$$\Delta S_{mix.} = -R\sum \kappa_i \ln \kappa_i \qquad (3.36)$$

In chemical reaction:

$$v_A A + v_B B \rightarrow v_M M + v_N N$$

the change of entropy is:

$$\Delta S_{chem.reac.} = (v_M S_M + v_N S_N) - (v_A S_A + v_B S_B) =$$

$$= \sum_{products}(v_j S_j) - \sum_{reac\tan ts}(v_i S_i) \qquad (3.37)$$

where v_i and v_j are the stoichiometric coefficients of reactants and products, but S_i and S_j represent the entropy values of compounds under given conditions.

The absolute value of entropy may be determined by the third law of thermodynamics. Early this law was known as Nernst's heat theorem:

$$\lim_{T \to 0} \Delta S = 0$$

Nowadays it is formulated as Planck's postulate:

$$S_0 = 0 \qquad (3.38)$$

or

$$\lim_{T \to 0} S = 0$$

Since entropy of a perfect crystalline substance is zero at zero Kelvin, then the value of entropy at temperature T may be represent as the change of entropy by heating of the substance from zero up to T K:

$$\Delta S = S_T - S_0 = S_T$$

Entropy change during heating from zero to T K is equal:

$$\Delta \overline{S}_{heat.} = \int_0^T \frac{\overline{c}_p}{T} dT \qquad (3.28)$$

Thus

$$\overline{S}_T = \int_0^T \frac{\overline{c}_p}{T} dT \qquad (3.39)$$

If in the range of $0 \div T$ K the phase transformation takes place, then (3.25) must be taken into account in the entropy expression:

$$\overline{S}_T = \int_0^{T_{fus.}} \frac{\overline{c}_p}{T} dT + \frac{\Delta \overline{H}_{fus.}}{T_{fus.}} + \int_{T_{fus.}}^{T_{boil.}} \frac{\overline{c}_p}{T} dT + \frac{\Delta \overline{H}_{boil.}}{T_{boil.}} + \int_{T_{boil.}}^{T} \frac{\overline{c}_p}{T} dT \qquad (3.40)$$

Let us return to entropy, as a criterion of spontaneity of the process. One of the formulations of the second law of thermodynamics runs as follows: "In isolated system a process proceeds in direction of entropy increase only". But in the nonisolated system during spontaneous process entropy may not only increase, but even decrease too e.g. at the crystallization and condensation $\Delta S_{syst.} < 0$. The entropy of a system is also reduced in the spontaneous process of the flattering of temperature, when $T_{syst.} > T_{surr.}$, etc. Thus, the entropy change of nonisolated system does not represent an ultimate criterion of spontaneity of process. The entropy change of so-called "universe" ("system in thermostat" according to other nomenclature) is presented in this role:

"Universe" = nonisolated system + surroundings.

According to the principle of additivity of entropy,

$$\Delta S_{"univ."} = \Delta S_{syst.} + \Delta S_{surr.} \tag{3.41}$$

It is accepted, that the entropy change of surroundings is stipulated by the heat exchange with the system only. Both in the reversible and irreversible processes:

$$q_{syst.} = -q_{surr.} \tag{3.42}$$

$$\frac{q_{syst.}}{T} = -\frac{q_{surr.}}{T} \tag{3.43}$$

$$\Delta S_{surr.} = \frac{q_{surr.}}{T} = -\frac{q_{syst.}}{T} \tag{3.44}$$

According to Clausius's well–known inequality, in spite of the system absorbs or releases the heat:

$$\Delta S_{syst.}^{irrev.} > \frac{q_{syst.}^{irrev.}}{T} \tag{3.22}$$

Let's write the expression (3.20) as equality:

$$\Delta S_{syst.} = \frac{q_{syst.}}{T} + \frac{q'}{T} \qquad (3.45)$$

where $q_{syst.}$ is the amount of heat, which exchanges between the system and surroundings, but q' represents so-called "uncompensated heat".

By Clausius,

$$q' \geq 0.$$

In the reversible process

$$\frac{q'}{T} = 0$$

and

$$\Delta S_{syst.}^{rev.} = \frac{q_{syst.}^{rev.}}{T} \qquad (3.21)$$

According to (3.44),

$$\Delta S_{surr.}^{rev.} = \frac{q_{surr.}^{rev.}}{T} = -\frac{q_{syst.}^{rev.}}{T} \qquad (3.44a)$$

By introducing the expressions (3.21) and (3.44a) into the equation (3.41), we obtain:

$$\Delta S_{"univ."}^{rev.} = \Delta S_{syst.}^{rev.} + \Delta S_{surr.}^{rev.} = \frac{q_{syst.}^{rev.}}{T} - \frac{q_{syst.}^{rev.}}{T} = 0 \qquad (3.46)$$

In the irreversible spontaneous process

$$\frac{q'}{T} > 0.$$

Then, according to the expressions (3.44) and (3.45),

$$\Delta S_{syst.}^{irrev.} = \frac{q_{syst.}^{irrev.}}{T} + \frac{q'}{T} \qquad (3.47)$$

$$\Delta S_{surr.}^{irrev.} = \frac{q_{surr.}^{irrev.}}{T} = -\frac{q_{syst.}^{irrev.}}{T} \qquad (3.44b)$$

and

$$\Delta S_{"univ."}^{irrev.} = \Delta S_{syst.}^{irrev.} + \Delta S_{surr.}^{irrev.} = \frac{q_{syst.}^{irrev.}}{T} + \frac{q'}{T} - \frac{q_{syst.}^{irrev.}}{T} = \frac{q'}{T} > 0 \qquad (3.48)$$

Thus, the spontaneous (natural) processes are characterized by the increasing of entropy of "universe". Entropy of nonisolated system may be decreased in natural process. But if this decrease is compensated with increase of entropy of surroundings and by so-called "noncompensated heat", the process is realized spontaneously.

It is necessary to consider here the reason that causes the occurrence of Clausius's inequality for irreversible (natural) processes:

$$\Delta S_{syst.}^{irrev.} > \frac{q_{syst.}^{irrev.}}{T} \qquad (3.22)$$

i.e. what is the meaning load of value q' in the expression (3.45)?

As is known, q represents the heat, received from surroundings or transferred to surroundings by the system. It is obvious, that the entropy changes in accordance with q. But what is q', which does not represent the heat exchange with surroundings (this heat is q), and in spite of this induces the increase of entropy?

The sole explanation of Clausius's inequality (3.22) and equality (3.45) consists in following: production of entropy takes place in irreversible process just because of irreversibility of the process itself.

Let write the expression (3.45) in the following form:

$$\Delta S_{syst.} = \Delta_e S + \Delta_i S \qquad (3.49)$$

where $\Delta_e S$ is the entropy change of the system, which is caused by the heat exchange with surroundings $(\Delta_e S = \frac{q}{T})$ and $\Delta_i S$ is the entropy change of the system, which is induced by the irreversibility of the process $(\Delta_i S = \frac{q'}{T})$.

The production of entropy does not take place in the reversible process:

$$\Delta_i S = 0 \qquad \text{and} \qquad \Delta S = \Delta_e S$$

In irreversible process the production of entropy occurs:

$$\Delta_i S > 0 \qquad \text{and} \qquad \Delta S = \Delta_e S + \Delta_i S$$

Besides, the more irreversible is the process, the more is the value of entropy, induced by the irreversibility of the process, e.g. in one of the extremely irreversible process (gas expansion in vacuum) the heat exchange with surroundings does not take place at all: $\Delta_e S = 0$ and entropy change is wholly stipulated by the "formation" of entropy: $\Delta S = \Delta_i S > 0$.

The comparison of the first and second laws of thermodynamics is interesting in relation to entropy "induction". The main point of the first law emphasizes that the energy is not formed from nothing and does not disappear without a trace:

$$\Delta_i U = 0 \qquad \text{and} \qquad \Delta U = \Delta_e U$$

But according to the second law:

$$\Delta_i S \geq 0 \qquad \text{and} \qquad \Delta S = \Delta_e S + \Delta_i S$$

This means that in the natural processes the entropy is formed as if from nothing. But entropy "induction" should cause the energy formation also, while $T\Delta S \sim \Delta U$. It follows from this that not only entropy but also energy are formed in the irreversible processes. This is not conformed to the law of energy constancy.

In fact, inside of system itself the redistribution of the energy resources takes place: that part of energy, by which the work (or ordered motion of particles) may be fulfilled, wholly or partially is transformed into the heat energy (or into the chaotic form of motion of particles). This results in production of entropy in the system $\Delta_i S > 0$ and in the wholly or partially loss of the ability of fulfillment of work by the system; so-called "lost work" is equal:

$$A_{lost} = T\Delta_i S = q' \ .$$

As a result of this redistribution the internal energy of the system remains constant.

The "noncompensated heat" q' does not represent heat indeed (physically), i.e. the amount of heat, which is exchanged between the system and surroundings. q' is that part of energy, which is consumed to fulfill work in the reversible process, but in the natural irreversible processes causes the same changes in the system, which should be induced by the receiving of the same amount of heat from surroundings. These changes are expressed in the occupation of new energy levels and in more uniform distribution of particles, i.e. in the increase of entropy.

Thus, in natural processes the system wholly or partially loses the ability to fulfill work. Besides of this, the energy dissipation occurs, which results in the degradation of a part of internal energy at the end of the process, i.e. in transformation of this part in the form, less available for us. It is often mentioned, that the danger of energy crisis exists in the world. By taking into consideration the abovementioned, this view may be more specified: the danger of not energy, but entopic crisis exists in the world. Nevertheless the total source of energy is constant, all processes proceeding in nature cause such distribution of this energy, that it becomes more and more difficult of access.

Problems

Problem № 3.1

Ludwig Boltzmann stated in his H-theorem (1872), that entropy of isolated system increases in spontaneous process and reaches maximum at the equilibrium. Many of scientists have considered, that H-theorem does not correspond to the laws of mechanics.

let us consider so-called "Zermelo's paradox". We assume, that system from initial state with entropy value $S_{init.}$, transfers spontaneously into the equilibrium state with entropy value $S_{equil.}$.

By Boltzmann's meaning,

$$S_{equil.} > S_{init.} \quad \text{and} \quad \Delta S_I = S_{equil.} - S_{init.} > 0 \,.$$

But according to Poincare's "recurrence" theorem, the finite mechanical system reverts again to the state, in which it resided formerly (Boltzmann agreed to this theorem also). Thus, the system from equilibrium state (with entropy $S_{equil.}$) transfers spontaneously into the initial state (with entropy $S_{init.}$). Entropy decreases in this process: $\Delta S_{II} = S_{init.} - S_{equil.} < 0$, since $S_{init.} < S_{equil.}$. Hence, both entropy increase and decrease are possible in the spontaneous processes. This means, that entropy does not represent quantity, by which the spontaneity of process may be characterized. Loschmidt's paradox was analogous.

Botzmann's answer on Zermelo's and Loschmidt's paradoxes and other similar "charges" was certain. What is Boltzmann's answer briefly?

Solution:

Boltzmann has answered so: law of entropy increase has not absolute but statistical character. If it is an absolute law, it would be in indubitable opposition with laws of mechanics. But the second law of thermodynamics due to its statistical character assumes the possibility of proceeding of processes with both entropy increase and entropy decrease. Another matter how probable are the spontaneous proceeding of those processes, which are characterized with entropy decrease (e.g. heat transfer from cold body to hot one, formation of temperature gradient from thermal equilibrium state, compression of gas, separation of mixture of gases into the constituents, etc).

Boltzmann has shown, that realization of these processes requires incomparably more time than duration of existence of universe. Boltzmann was correct, when he answered to Zermelo: you will have to wait very long for the supervision the spontaneous proceeding of the similar (i.e. unnatural) processes. In next problems we shall see, what time is required for the approach of so-called "Great fluctuation" i.e. spontaneous transferring of the system from equilibrium state into the state strongly removed from it.

Problem № 3.2

Two closed thermodynamic systems are given. Each system contains 6×10^{22} atoms of helium. Temperatures of the first and second system are equal to 500 K and 300 K respectively. Systems were got into thermal contact, proceeded until the thermal equilibrium state is reached. Equilibrium temperature of the combined system is 400 K.

What are the values of $\Delta\sigma$ and ΔS of this process?

Solution:

$\Delta\sigma$ of the process is equal to the change of σ of combined system (hot system + cold system) due to heat transfer. σ is an additive quantity and therefore $\Delta\sigma$ of the combined system is equal to the sum of $\Delta\sigma$ of its constituting systems:

$$\Delta\sigma_{comb.} = \Delta\sigma_h + \Delta\sigma_c \qquad (a)$$

where $\Delta\sigma_h$ and $\Delta\sigma_c$ are changes of σ of hoter and colder systems respectively at heat transfer.

Let us use Sacure—Tetrode's expression for calculation of $\Delta\sigma$:

$$\sigma = N\left(\ln T^{3/2} - \ln \frac{N}{V} = const \right) \qquad (3.10)$$

Number of particles and the volume of hot and cold systems do not change at heat transfer. Taking into account this condition it may be obtained from expression (3.10):

$$\Delta\sigma = \frac{3}{2} N \ln \frac{T_{equil.}}{T_{init.}} \qquad (b)$$

According to (b)

$$\Delta\sigma_h = \frac{3}{2} N \ln \frac{T_{equil.}}{T_h}$$

$$(c)$$

$$\Delta\sigma_c = \frac{3}{2} N \ln \frac{T_{equil.}}{T_c}$$

Let us introduce expressions (c) into (a):

$$\Delta\sigma_{comb.} = \frac{3}{2} N\left(\ln \frac{T_{equil.}}{T_h} + \ln \frac{T_{equil.}}{T_c} \right) = 1.5 \times 6 \times 10^{22} \times \left(\ln \frac{400}{500} + \ln \frac{400}{300} \right) =$$

$$= 9 \times 10^{22} (-0.223 + 0.288) = 5.81 \times 10^{21}$$

According to expression (3.9):

$$\Delta S = k\Delta\sigma = 1.38 \times 10^{-23} \times 5.81 \times 10^{21} = 8.01 \times 10^{-2} = 0.08 J / K$$

Problem № 3.3

Two closed thermodynamic systems are in thermal equilibrium with each other ($T_{equil.}$ = 400 K). Each system contains 6×10^{22} atoms of helium. What time should be passed till the occurrence of so-called "Great

Fluctuation", when the systems should be found in their initial state i.e. when temperatures of the first and second systems will become 500 K and 300 K respectively?

Solution:

The value of $\Delta\sigma$ was calculated in problem № 3.2 for the reverse process – transfer of two systems with different temperatures (T_{hot} = 500 K and T_{cold} = 300 K) into the thermal equilibrium state ($T_{equil.}$ = 400 K). It was obtained, that for combined system:

$$\Delta\sigma_{comb.syst.} = 5.81 \times 10^{21}$$

As is known,

$$\Delta\sigma_{comb.syst.} = \ln\frac{W_{equl.}}{W_{init.}}$$

That is,

$$\ln\frac{W_{equil.}}{W_{init.}} = 5.81 \times 10^{21}$$

hence

$$\lg\frac{W_{equil.}}{W_{init.}} = \frac{5.81 \times 10^{21}}{2.3} = 2.53 \times 10^{21}$$

and

$$\frac{W_{equil.}}{W_{init.}} \approx 10^{10^{21}}$$

$W_{equil.}$ is the number of microstates, by means of the thermal equilibrium is realized, but $W_{init.}$ represents the number of microstates, by which the initial state is performed. Since the probability of the equilibrium state approximately $10^{10^{21}}$ – fold exceeds the probability of the initial state, therefore in the combined system, which is in the thermal equilibrium state, the number of microstates should be changed approximately $10^{10^{21}}$ – fold, until it's reversal in the initial state (i.e. the formation of temperature difference, pointed out in the condition of problem) becomes possible.

Let us see, what time is required for this. Let consider, that the microstate of system changes after each collision of atoms. Let assume, that one atom undergoes 10^{10} collision per second. Then $2 \times 6 \times 10^{22}$ atoms, presented in the combined system, undergo $12 \times 10^{22} \times 10^{10} = 1.2 \times 10^{33}$ collisions in second, but $1.2 \times 10^{33} \times 3600 \times 24 \times 365 = 3.78 \times 10^{40}$ collisions in a year. Thus, the microstate of the combined system is varied 3.78×10^{40} – fold in a year. The approach of "Great Fluctuation", or the change of microstate of system $10^{10^{21}}$ – fold requires:

$$\frac{10^{10^{21}}}{3.8 \times 10^{40}} \quad \text{years.}$$

In order to estimate the value of this expression clearly, let us consider its logarithm:

$$\lg \frac{10^{10^{21}}}{3.8 \times 10^{40}} = 10^{21} - 40.6$$

40.6 is incomparably minute with 10^{21}. So it may be accepted, that $10^{21} - 40.6 \approx 10^{21}$ and

$$\frac{10^{10^{21}}}{3.8 \times 10^{40}} \approx 10^{10^{21}} \quad \text{years}$$

Thus, the occurrence of "Great fluctuation" requires approximately $10^{10^{21}}$ years. If take into account, that the average duration of human life is equal to 65 years and duration of the world existence numbers approximately 10^{10} years, then it is obvious, that the "Great fluctuation" is never occurred from the beginning of the world existence till nowadays and it will never take place practically.

So the temperature difference will never be formed without external influence in the system, presented in thermal equilibrium.

Remark: It is shown clearly from this example, why the second law of thermodynamics (or the law of entropy increase in spontaneous processes) has a statistical character. This law is statistical and not absolute, since assumes the possibility of spontaneous procceding of process in direction of both increase and decrease of entropy. Neverthereless, it is the law, because the possibility of a process with decreasing of entropy is negligible. Practically, the proceeding of such process is excluded.

Problem № 3.4

26 g xenon is placed in the container under 2 atm pressure and at ambient temperature. The system expands isothermally until 1 atm pressure.

What are the values of $\Delta\sigma$ and ΔS of process?

Solution:

The system is in equilibrium state before the beginning of expansion (as well as at the end of expansion). Its thermodynamic probability W_1 is expressed by equation (3.3):

$$W_1 = \frac{N!}{\left(\left[\frac{N}{r}\right]!\right)^r}$$

where N is number of particles in the system, and r represents number of energy levels.

According to condition of problem, pressure of the system reduces two – fold due to isothermal expansion. This means, that volume of system increases two–fold. The number of energy levels in this new volume will be twice more than it is before expansion. The thermodynamic probability of the system after expansion is:

$$W_2 = \frac{N!}{\left(\left[\dfrac{N}{2r}\right]!\right)^{2r}}$$

According to expression (3.7):

$$\Delta\sigma = \ln\frac{W_2}{W_1} = \ln W_2 - \ln W_1$$

Let us use Stirling's equation for the calculation of factorials:

$$\ln N! = N \ln N - N \qquad (3.4)$$

$$\ln W_2 = \ln \frac{N!}{\left(\left[\dfrac{N}{2r}\right]!\right)^{2r}} = \ln N! - 2r\ln\left(\frac{N}{2r}\right)! = N\ln N - N - 2r\left(\frac{N}{2r}\ln\frac{N}{2r} - \frac{N}{2r}\right) =$$

$$= N\ln N - N - N\ln\frac{N}{2r} + N = N\left(\ln N - \ln\frac{N}{2r}\right) = N\ln 2r$$

$$\ln W_1 = \ln\frac{N!}{\left(\left[\dfrac{N}{r}\right]!\right)^{r}} = \ln N! - r\ln\left(\frac{N}{r}\right)! = N\ln N - N - r\left(\frac{N}{r}\ln\frac{N}{r} - \frac{N}{r}\right) =$$

$$= N \ln N - N - N \ln \frac{N}{r} + N = N\left(\ln N - \ln \frac{N}{r}\right) = N \ln r$$

Thus,

$$\Delta\sigma = \ln W_2 - \ln W_1 = N \ln 2r - N \ln r = N \ln 2$$

According to condition of problem, 26 *g* or 0.2 *M* of xenon is present in the system.

$$N = 0.2 \times 6 \times 10^{23} = 1.2 \times 10^{23} \ atoms.$$

$$\Delta\sigma = N \ln2 = 1.2 \times 10^{23} \times 0.693 = 0.83 \times 10^{23}$$

Following expression (3.9),

$$\Delta S = k\Delta\sigma = 1.38 \times 10^{-23} \times 0.83 \times 10^{23} = 1.15 \ J/K$$

Problem № 3.5

1.2×10^{23} atoms of xenon is present in the container under 1 atm pressure and at ambient temperature. What time should be passed until the approach of "Great fluctuation" i.e. until spontaneous reducing of gas volume by two–fold?

Solution:

It was calculated in Problem № 3.4, that by two–fold increasing of volume of 1.2×10^{23} atoms xenon in isothermal conditions:

$$\Delta\sigma = \ln \frac{W_2}{W_1} = 0.83 \times 10^{23}$$

where W_1 and W_2 are numbers of microstates before and after expansion respectively. Hence

$$\lg \frac{W_2}{W_1} = \frac{0.83 \times 10^{23}}{2.3} = 3.61 \times 10^{22}$$

and

$$\frac{W_2}{W_1} \approx 10^{10^{21}}$$

Thus, number of the macrostates must be changed $10^{10^{22}}$ – fold in the expanded system, until spontaneous two–fold reducing of volume of the system becomes possible (more scrupulous see problem № 3.3).

Let us assume that each atom undergoes ~ 10^{10} collisions per second. Then 1.2×10^{23} atoms undergo in a year:

$$1.2 \times 10^{23} \times 10^{10} \times 3600 \times 24 \times 365 = 3.78 \times 10^{40} \text{ collisions.}$$

If consider, that microstate of system changes after each collision, then the approach of great fluctuation (or spontaneous compression of system) needs:

$$\frac{10^{10^{22}}}{3.8 \times 10^{40}} \approx 10^{10^{22}} \quad \text{years}$$

This is so much time (duration of existence of the universe is ~ 10^{10} years), that it should be mentioned boldly: the spontaneous compression of gas did not never take place and will never occur. The analogous result is obtained in case of spontaneous creation of temperature difference from thermal equilibrium state (see problem № 3.3).

Thus, the probability of spontaneous reverting of system from equilibrium state to the initial state is practically equal to zero.

Problem № 3.6

In problems №№ 3.3 and 3.5 we have seen, that the spontaneous transfer of the system from equilibrium state into the state, being tremendously distant from equilibrium (e.g. initial) is practically excluded. This does not mean however that the less probable states than equilibrium one are not realized in general. They are accomplished and sufficiently often, but especially those distributions, probability of which is not distinguished considerably from the probability of equilibrium state.

The transfer of the system from the most probable state (from equilibrium state) into the less one is referred to as fluctuation. The equilibrium between system and surroundings is disturbed due to fluctuation i.e. the equality of the characterizing parameters of system and surroundings is disrupted. If at equilibrium $T_{syst.} = T_{surr.}$, then because of temperature fluctuation $T_{syst.} - T_{surr.} = \Delta T$. The average squared dispersion of temperature (fluctuation) is expressed so:

$$\sqrt{\overline{(\Delta T)^2}} = \sqrt{\frac{k}{c_v}} \cdot T_{surr.} \tag{3.50}$$

where $\overline{(\Delta T)^2} = (T - \overline{T})^2$, k is Boltzmann's constant.

The relative fluctuation of temperature is equal:

$$\delta_T = \frac{\sqrt{\overline{(\Delta T)^2}}}{T_{surr.}} = \sqrt{\frac{k}{c_v}} \tag{3.51}$$

1) Deduce the expressions of the average and relative fluctuation for moderate and high temperatures.

2) Determine the value of relative fluctuation of temperature of 1 and 10^{-5} mole monoatomic ideal gas at 300, 500, 1000 K.

Solution:

1) The heat capacity of monoatomic ideal gas at moderate and high temperatures is equal:

$$c_v = 3/2 \ Rn = 3/2 \ kN \tag{3.52}$$

where n is quantity of moles, N represents number of particles, R is gas constant and k – Boltzmann's constant.

Introduce the expression (3.52) into the (3.50) and (3.51):

$$\sqrt{\overline{(\Delta T)^2}} = \sqrt{\frac{k}{c_v}} \cdot T_{surr.} = \sqrt{\frac{k}{\frac{3}{2} Nk}} \cdot T_{surr.} = \sqrt{\frac{2}{3N}} \cdot T_{surr.} \tag{3.53}$$

$$\delta_T = \frac{\sqrt{\frac{2}{3N}}}{T_{surr.}} \cdot T_{surr.} = \sqrt{\frac{2}{3N}} \tag{3.54}$$

2) $6 \cdot \times 10^{23}$ *atoms* are in one *mole* of ideal gas. The relative fluctuation is equal:

$$\delta_T = \sqrt{\frac{2}{3N}} = \sqrt{\frac{2}{3 \times 6 \times 10^{23}}} = 1.05 \times 10^{-12}$$

10^{-5} *mole* of ideal gas contains 6×10^{18} *atoms*:

$$\delta_T = \sqrt{\frac{2}{3N}} = \sqrt{\frac{2}{3 \times 6 \times 10^{18}}} = 3.33 \times 10^{-10}$$

Thus, the reducing of number of particles causes the increase of relative fluctuation. It should be so, while the second law of thermodynamics possesses the statistical character: the more the number of particles, the more is the probability of fulfillment of this law and the less is the value of fluctuation.

Problem № 3.7

Gaseous thermometer contains 10^{-5} mole of ideal gas. What is the precision limit of this instrument in the range $300 \div 1000$ K?

Solution:

The precision limit of thermometer is conditioned by fluctuation of temperature, which takes place in the measuring instrument. The fluctuation of temperature from comparison of expressions (3.50) and (3.51) is equal to:

$$\sqrt{\overline{(\Delta T)^2}} = \delta_T \times T \tag{3.55}$$

In the problem № 3.6 we have determined, that the relative fluctuation of temperature of 10^{-5} *mole* ideal gas is $\delta_T = 3.33 \times 10^{-10}$. Hence,

$$\sqrt{\overline{(\Delta T)^2}} = 3.33 \times 10^{-10} \times T$$

300 K $\sqrt{\overline{(\Delta T)^2}} = 3.33 \times 10^{-10} \times 300 = 1 \times 10^{-7}$ *degree*

500 K $\sqrt{\overline{(\Delta T)^2}} = 3.33 \times 10^{-10} \times 500 = 1.67 \times 10^{-7}$ *degree*

1000 K $\sqrt{\overline{(\Delta T)^2}} = 3.33 \times 10^{-10} \times 1000 = 3.33 \times 10^{-7}$ *degree*

Thus, the precision limit of gaseous thermometer in the range $300 \div 1000$ K is $\sim 10^{-7}$ *degree*. When the temperature change fixed by gaseous thermometer $\leq 10^{-7}$ *degree*, we will not understand the reason of this temperature change. It may be stipulated by temperature fluctuation of thermometer or by any process which takes place in the system.

Problem № 3.8

The general expression for relative fluctuation of energy is following:

$$\delta_E = \frac{T\sqrt{kc_v}}{E} \tag{3.56}$$

1) Derive the expression for relative fluctuation of energy at moderate and high temperatures.

2) Determine the value of δ_E for 1 mole and 10^{-5} mole of ideal gas at moderate and high temperatures.

Solution:

1) The heat capacity and energy of ideal gas at moderate and high temperatures are expressed so:

$$C_v = \frac{3}{2} N k \qquad (3.52)$$

$$E = \frac{3}{2} NkT \qquad (3.57)$$

Introduce the expressions (3.52) and (3.57) into the (3.56):

$$\delta_E = \frac{T\sqrt{kc_v}}{E} = \frac{T\sqrt{\frac{3}{2}Nk^2}}{\frac{3}{2}NkT} = \frac{k\sqrt{\frac{3}{2}N}}{\frac{3}{2}Nk} = \frac{1}{\sqrt{\frac{3}{2}N}} = \sqrt{\frac{2}{3N}} \qquad (3.58)$$

2) $N = 6 \times 10^{23}$ *atoms.* $\qquad \delta_E = \sqrt{\dfrac{2}{3 \times 6 \times 10^{23}}} = 1.05 \times 10^{-12}$

$N = 6 \times 10^{18}$ *atoms.* $\qquad \delta_E = \sqrt{\dfrac{2}{3 \times 6 \times 10^{18}}} = 3.33 \times 10^{-10}$

Thus, the value of relative fluctuation of energy is increased with the decrease of particles' number. The same result is obtained in the case of temperature fluctuation (problem № 3.6). It is interesting to note, that the relative fluctuations of energy and temperature have the same values:

$$\delta_E = \delta_T = \sqrt{\frac{2}{3N}}$$

Problem № 3.9

Debye's law of T^3 describes the dependence of heat capacity on the temperature at the very law temperatures:

$$C_V = 2{,}5\pi^4 Nk \left(\frac{T}{\Theta}\right)^3 \qquad (3.59)$$

The system energy is equal:

$$E = 0{,}6\pi^4 NkT \left(\frac{T}{\Theta}\right)^3 \qquad (3.60)$$

where Θ is Debye characteristic temperature. *

Derive the formulae of relative fluctuation of temperature and energy for the very law temperatures.

Solution:

The relative fluctuation of temperature is generally expressed so:

$$\delta_T = \sqrt{\frac{k}{c_v}} \qquad (3.51)$$

Introduce in expression (3.51) the equation (3.59) for temperature dependence of heat capacity at low temperatures:

$$\delta_T = \sqrt{\frac{k}{c_v}} = \sqrt{\frac{k}{2.5\pi^4 Nk\left(\dfrac{T}{\Theta}\right)^3}} = \sqrt{\frac{1}{2.5\pi^4}} \sqrt{\frac{1}{N}\left(\frac{\Theta}{T}\right)^3} = 0.064\sqrt{\frac{1}{N}\left(\frac{\Theta}{T}\right)^3}$$

$$(3.61)$$

---

*do not entangle Debye characteristic temperature (Θ) with vibrational characteristic temperature $(\Theta_{vibr.})$, which is applied in Einstein's functions.

Introduce also in the general formula of δ_E (3.58) the expressions (3.59) and (3.60) for heat capacity and energy:

$$\delta_E = \frac{T\sqrt{kc_v}}{E} = \frac{T\sqrt{k \times 2.5\pi^4 Nk \left(\dfrac{T}{\Theta}\right)^3}}{0.6\pi^4 NkT \left(\dfrac{T}{\Theta}\right)^3} = \frac{k\pi^2 \sqrt{2.5N \left(\dfrac{T}{\Theta}\right)^3}}{k\pi^4 \cdot 0.6N \left(\dfrac{T}{\Theta}\right)^3} =$$

$$= \frac{\sqrt{2.5}}{0.6\pi^2} \cdot \frac{1}{\sqrt{N \cdot \left(\dfrac{T}{\Theta}\right)^3}} = 0.267 \sqrt{\frac{1}{N} \left(\frac{\Theta}{T}\right)^3} \qquad (3.62)$$

Problem № 3.10

Use the expressions for δ_T (3.61) and δ_E (3.62), obtained in problem № 3.9 and determine the relative fluctuations of temperature and energy at 10^{-2} and $10^{-5} K$ for a mole of argon ($\Theta_{Ar} = 92$).

Tabulate the data on the basis of results obtained in problems №3.6, №3.8 and №3.10.

Solution

1) $T = 10^{-2} K$; $\quad \dfrac{\Theta}{T} = \dfrac{92}{10^{-2}} = 92 \cdot 10^2$; $\qquad \left(\dfrac{\Theta}{T}\right)^3 = 7{,}79 \cdot 10^{11}$

According to the expressions (3.61) and (3.62),

$$\delta_T = 0.064 \sqrt{\frac{1}{N} \left(\frac{\Theta}{T}\right)^3} \cdot = 0.064 \sqrt{\frac{7.79 \times 10^{11}}{6 \times 10^{23}}} = 0.73 \times 10^{-7}$$

$$\delta_E = 0.267 \sqrt{\frac{1}{N} \left(\frac{\Theta}{T}\right)^3} = 0.267 \sqrt{\frac{7.79 \times 10^{11}}{6 \times 10^{23}}} = 3.04 \times 10^{-7}$$

$$2)\,T = 10^{-5}\,K\,;\quad \frac{\Theta}{T} = \frac{92}{10^{-5}} = 92 \cdot 10^{5}\,;\qquad\qquad \left(\frac{\Theta}{T}\right)^{3} = 7.79 \times 10^{20}$$

$$\delta_{T} = 0.064\sqrt{\frac{1}{N}\left(\frac{\Theta}{T}\right)^{3}} = 0.064\sqrt{\frac{7.79 \times 10^{20}}{6 \times 10^{23}}} = 0.23 \times 10^{-2}$$

$$\delta_{E} = 0.267\sqrt{\frac{1}{N}\left(\frac{\Theta}{T}\right)^{3}} = 0.267\sqrt{\frac{7.79 \times 10^{20}}{6 \times 10^{23}}} = 0.96 \times 10^{-2}$$

Let enter the results of problems №3.6, № 3.8 and № 3.10 in the table.

Table 3.1. Dependence of relative fluctuations of temperature and energy on temperature $(N_{Ar} = 6 \times 10^{23}$ atoms).

Temperature, K	δ_T	δ_E
300 ÷ 1000	1.05×10^{-12}	1.05×10^{-12}
10^{-2}	0.73×10^{-7}	3.04×10^{-7}
10^{-5}	0.23×10^{-2}	0.96×10^{-2}

As is shown from Table 3.1, δ_T and δ_E do not represent temperature function at middle and high temperatures (these values depend on the number of particles only). But near 0 K the fluctuation is characterized by considerable temperature dependence, in particular, the values of fluctuation sharply increase with reduction of temperature. Such difference is caused by the reason that at middle and high temperatures $c_V \neq f(T)$ and $E \sim T^1$, but at very low temperatures $c_V \sim T^3$ and $E \sim T^4$.

Problem № 3.11

If gas is present in classical regime, it is ideal; if average number of atoms on the orbital is much less than one, regime hold classical. In the classical conditions $CV_Q \ll 1$, where C is concentration of gas, but V_Q represents quantum volume of particle: $V_Q = \left(\dfrac{h^2}{2\pi mkT} \right)^{3/2}$.

Estimate, if $1L$ of helium represents an ideal gas in the following conditions: 1) 5K, 150 atm; 2) 10K, 1000 atm; 3)100K, 1 atm; 4) 300K, 1 atm; 5) 500K, 100 atm. Tabulate the obtained results.

Solution:

Let us give the value of V_Q in CGSM units.

$h = 6.626 \times 10^{-27}\ erg \times second$ (Planck's constant)
$k = 1.38 \times 10^{-16}\ erg / K$ (Boltzmann's constant)
$m_{He} = 4 \times 1.66 \times 10^{-24} = 6.64 \times 10^{-24}\ g$

$$
V_Q = \left(\frac{h^2}{2\pi mkT} \right)^{3/2} = \left[\frac{(6.626 \times 10^{-27})^2\ erg^2 \times s^2}{6.28 \times 6.64 \times 10^{-24}\ g \times 1.38 \times 10^{-16}\ \frac{erg}{K} \times TK} \right]^{3/2} =
$$

$$
= \left[\frac{0.763 \times 10^{-14}}{T}\ \frac{erg \times s^2}{g} \right]^{3/2} = \left[\frac{0.763 \times 10^{-14}}{T}\ \frac{g \times cm^2}{s^2}\ \frac{s^2}{g} \right]^{3/2} = \qquad \text{(a)}
$$

$$
= \left[\frac{0.763 \times 10^{-14}}{T}\ cm^2 \right]^{3/2} = \frac{0.666 \times 10^{-21}}{T^{3/2}}\ cm^3
$$

Let us give *concentration* (C) with *atom/cm³*. From the ideal gas equation:

$$
C = \frac{nN_A}{V} = \frac{PN_A}{RT}
$$

In conditions of problem we have P and T. Therefore the expression $C = \dfrac{PN_A}{RT}$ is used. Because P is given in *atm*, for R we will use:

$$R = 0.082 \frac{L \times atm}{K \times mol}. \text{ Then}$$

$$C = \frac{6 \times 10^{23}}{0.082} \times \frac{P}{T} \frac{\dfrac{atom}{mol}}{\dfrac{L \times atm}{K \times mol}} \times \frac{atm}{K} = 7.317 \times 10^{24} \times \frac{P}{T} \frac{atom}{L} =$$

$$= \frac{7.317 \times 10^{24}}{10^3} \times \frac{P}{T} \frac{atom}{cm^3} = 7.317 \times 10^{21} \times \frac{P}{T} \frac{atom}{cm^3}$$

(b)

(where P is expressed by *atmospheres*).

Let us introduce the values of T and P into expressions (a) and (b):

1) $T = 5$ K, $P = 150$ *atm*.

$$C = 7.317 \times 10^{21} \times \frac{150}{5} = 2.2 \times 10^{23} \frac{atom}{cm^3}$$

$$V_Q = \frac{0.666 \times 10^{-21}}{5^{3/2}} = \frac{0.666 \times 10^{-21}}{11.18} = 5.96 \times 10^{-23} \ cm^3$$

$$CV_Q = 2.2 \times 10^{23} \times 5.96 \times 10^{-23} = 13.09 \ atoms.$$

$CV_Q > 1$ i.e. helium is in quantum regime. Therefore helium does not represent an ideal gas in the given conditions (5 K, 150 *atm*).

Remark. CV_Q indicates the number of atoms, which exists in the quantum volume (V_Q) of helium at the given conditions. This amount of atoms is closely related with average population of orbital. When $CV_Q > 1$, there are more than one atom on the orbital and conditions are not classical already i.e. gas is not ideal.

2) $T = 10$ K, $P = 1000$ *atm*.

$$C = 7.317 \times 10^{21} \times \frac{P}{T} = 7.317 \times 10^{21} \times \frac{10^3}{10} = 7.32 \times 10^{23} \frac{atom}{cm^3}$$

$$V_Q = \frac{0.666 \times 10^{-21}}{T^{3/2}} = \frac{0.666 \times 10^{-21}}{10^{3/2}} = 2.11 \times 10^{-23} \; cm^3$$

$$CV_Q = 7.32 \times 10^{23} \times 2.11 \times 10^{-23} = 15.4 \; atoms$$

$CV_Q > 1$, i.e. helium does not present in the state of ideal gas under given conditions.

3) $T = 100$ K, $P = 1$ atm.

$$C = 7.32 \times 10^{21} \times \frac{1}{100} = 7.32 \times 10^{19} \frac{atom}{cm^3}$$

$$V_Q = \frac{0.666 \times 10^{-21}}{100^{3/2}} = 6.66 \times 10^{-25} \; cm^3$$

$$CV_Q = 7.32 \times 10^{19} \times 6.66 \times 10^{-25} = 0.49 \times 10^{-4} \; atoms$$

$CV_Q \ll 1$, thus helium is ideal gas under the given conditions.

4) $T = 300$ K, $P = 1$ atm.

$$C = 7.32 \times 10^{21} \times \frac{1}{300} = 2.44 \times 10^{19} \frac{atom}{cm^3}$$

$$V_Q = \frac{0.666 \times 10^{-21}}{300^{3/2}} = 1.28 \times 10^{-25} \; cm^3$$

$$CV_Q = 2.44 \times 10^{19} \times 1.28 \times 10^{-25} = 0.31 \times 10^{-5} \; atoms$$

$CV_Q \ll 1$, i.e. helium is in the state of ideal gas under given conditions.

5) $T = 500$ K, $P = 100$ atm.

$$C = 7.32 \times 10^{21} \times \frac{100}{500} = 1.46 \times 10^{21} \quad \frac{atom}{cm^3}$$

$$V_Q = \frac{0.666 \times 10^{-21}}{500^{3/2}} = 5.96 \times 10^{-26} \quad cm^3$$

$$CV_Q = 1.46 \times 10^{21} \times 5.96 \times 10^{-26} = 0.87 \times 10^{-4} \quad atoms$$

$CV_Q \ll 1$, helium exists an ideal gas under the given conditions.

Table 3.2. Results of problem № 3.11

T K	P, atm	C, atom/cm^3	V_{Q}, cm^3	CV_Q, atom	Regime	Gas
5	150	2.2×10^{23}	6×10^{-23}	13.1	Quantum	Nonideal
10	1000	7.3×10^{23}	2×10^{-23}	15.4	Quantum	Nonideal
100	1	7.3×10^{19}	7×10^{-25}	5×10^{-5}	Classical	Ideal
300	1	2.4×10^{19}	1×10^{-25}	0.3×10^{-5}	Classical	Ideal
500	100	1.5×10^{21}	0.6×10^{-25}	8.7×10^{-5}	Classical	Ideal

Problem № 3.12

Entropy of monoatomic ideal gas is expressed so:

$$S = kN\left(\frac{3}{2}\ln T - \ln\frac{N}{V} + const\right) \qquad (3.11)$$

where

$$const = \ln\left(\frac{2\pi mk}{h^2}\right)^{3/2} + \frac{5}{2} \qquad (3.12)$$

Determine the numerical value of *constant* for helium in:
a) CGSM, b) SI units.

Solution:

a) In CGSM units:

$h = 6.626 \times 10^{-27}$ *erg* $\times$ *second*
$k = 1.38 \times 10^{-16}$ *erg* $/ K$
$m_{He} = 4.003 \times 1.66 \times 10^{-24}$ $g = 6.64 \times 10^{-24}$ g

$$\frac{2\pi mk}{h^2} = \frac{(6.28) \times (6.64 \times 10^{-24}\ g) \times (1.38 \times 10^{-16}\ erg \times K^{-1})}{(6.626 \times 10^{-27}\ erg \times s)^2} =$$

$$= 1.31 \times 10^{14}\ \frac{g \times K^{-1}}{erg \times s^2} = 1.31 \times 10^{14}\ cm^{-2} \times K^{-1}$$

$$\left(1 erg = 1\ \frac{g \times cm^2}{s^2} \right)$$

$$const = \frac{3}{2} \ln \left(\frac{2\pi mk}{h^2} \right) + \frac{5}{2} = \frac{3}{2} \ln 1.31 \times 10^{14} + \frac{5}{2} = 48.76 + 2.5 = 51.26$$

b) In SI units:

$h = 6.626 \times 10^{-34}$ $J \times s$
$k = 1.38 \times 10^{-23}$ J / K
$m_{He} = 6.64 \times 10^{-27}$ kg

$$\frac{2\pi mk}{h^2} = \frac{(6.28) \times (6.64 \times 10^{-27}\ kg) \times (1.38 \times 10^{-23}\ J \times K^{-1})}{(6.626 \times 10^{-34}\ J \times s)^2} =$$

$$= 1.31 \times 10^{18}\ \frac{kg \times K^{-1}}{J \times s^2} = 1.31 \times 10^{18}\ m^{-2} \times K^{-1}$$

$$\left(1 J = 1\ \frac{kg \times m^2}{s^2} \right)$$

$$const = \frac{3}{2}\ln 1.31 \times 10^{18} + \frac{5}{2} = 62.57 + 2.5 = 65.07$$

Problem № 3.13

1 L of helium is present under 1 atm pressure and an ambient temperature. Determine its σ and S in: a) CGSM and b) SI units.

Solution:

We use expressions:

$$\sigma = N\left(\frac{3}{2}\ln T - \ln\frac{N}{V} + const\right) \tag{3.10}$$

$$S = k\sigma \tag{3.8}$$

$$N = nN_A \tag{a}$$

where n is number of moles, N represents number of particles in the system and N_A is Avogadro's number.

$PV = nRT$ according to ideal gas equation. From expression (a) it is obtained:

$$N = \frac{PV}{RT} \times N_A = \frac{1 \times 1 \times 6 \times 10^{23}}{0.082 \times 298} = 0.25 \times 10^{23} \text{ } atoms$$

a) In CGSM units:

$V = 1\,L = 10^3\,cm^3$

$const = 51.26$ (see problem № 3.12), then

$$\sigma = N\left(\frac{3}{2}\ln T + \ln V - \ln N + const\right) = 0.25 \times 10^{23} \times$$

$$\times \left(\frac{3}{2}\ln 298 + \ln 1 \times 10^3 - \ln 0.25 \times 10^{23} + 51.26\right) =$$

$$= 0.25 \times 10^{23}(8.55 + 6.91 - 51.50 + 51.26) = 0.25 \times 10^{23} \times 15.15 = 3.79 \times 10^{23}$$

$$S = k\sigma = 1.38 \times 10^{-16} \times 3.79 \times 10^{23} = 5.23 \times 10^{7} \ erg \, / \, K$$

b) In SI units:

$$V = 1 \ L = 10^{-3} \ m^3$$

const = 65.07 (see problem № 3.12), hence

$$\sigma = N\left(\frac{3}{2}\ln T + \ln V - \ln N + const\right) = 0.25 \times 10^{23} \times$$

$$\times\left(\frac{3}{2}\ln 298 + \ln 1 \times 10^{-3} - \ln 0.25 \times 10^{23} + 65.07\right) =$$

$$= 0.25 \times 10^{23}(8.55 - 6.91 - 51.57 + 65.07) = 0.25 \times 10^{23} \times 15.13 = 3.78 \times 10^{23}$$

$$S = k\sigma = 1.38 \times 10^{-23} \times 3.78 \times 10^{23} = 5.22 \ J \, / \, K$$

Remark. σ has the same value in both systems of units. (σ is dimensionless quantity). The values of S are different in CGSM and SI units, which is conditioned by various values of Boltzmann's constant in these systems.

Problem № 3.14
Show by using of Sacure—Tetrode's formula (3.10), that $\sigma = f(N,T,P)$. Determine the numerical value of *constant* for helium in CGSM and SI units.

Solution:

$$\sigma = N\left(\frac{3}{2}\ln T - \ln\frac{N}{V} + const\right) \tag{3.10}$$

From ideal gas equation:

$$\frac{N}{V} = \frac{P}{kT} \qquad\qquad (a)$$

Let us introduce the value of N/V from (a) into (3.10). We obtain:

$$\sigma = N\left(\frac{3}{2}\ln T - \ln\frac{P}{kT} + const\right) \qquad\qquad (3.63)$$

Let us convert expression (3.63):

$$\sigma = N\left(\frac{5}{2}\ln T - \ln P + \ln k + const\right)$$

We designate:

$$\ln k + const = const' \qquad\qquad (b)$$

Then

$$\sigma = N\left(\frac{5}{2}\ln T - \ln P + const'\right) \qquad\qquad (3.64)$$

and

$$S = kN\left(\frac{5}{2}\ln T - \ln P + const'\right) \qquad\qquad (3.65)$$

where

$$const' = \ln\left(\frac{2\pi mk}{h^2}\right)^{3/2} + \frac{5}{2} + \ln k \qquad\qquad (3.66)$$

According to expressions (3.64) and (3.65), both σ and S represent function of N, T and P.

Let us determine the numerical value of $const\,' = const + \ln k$ for helium:

a) In CGSM units:

$k = 1.38 \times 10^{-16}\ erg/K;\quad const = 51.26$ (see problem № 3.12), then

$const\,' = \ln k + const = \ln 1.38 \times 10^{-16} + 51.26 = -36.52 + 51.26 = 14.74$

b) In SI units:

$k = 1.38 \times 10^{-23}\ J/K;\quad const = 65.07$ (see problem № 3.12),

$const\,' = \ln k + const = \ln 1.38 \times 10^{-23} + 65.07 = -52.64 + 65.07 = 12.43$

Tabulate the values of constants in CGSM and SI units.

Table 3.3. The values of constants in Sacure—Tetrode's expression for helium

Entropy expression	Expression for constant	Numerical value of constant CGSM	SI
$S = Nk\left(\dfrac{3}{2}\ln T - \ln\dfrac{N}{V} + const\right)$ (3.11)	$const = \ln\left(\dfrac{2\pi mk}{h^2}\right)^{3/2} + \dfrac{5}{2}$ (3.12)	51.26	65.07
$S = Nk\left(\dfrac{5}{2}\ln T - \ln P + const'\right)$ (3.65)	$const' = \ln\left(\dfrac{2\pi mk}{h^2}\right)^{3/2} + \dfrac{5}{2} + \ln k$ (3.66)	14.74	12.43

It should be noted, that constant in Sacure—Tetrode's equation represents so–called chemical constant of gas. Deducing of its expression was impossible by classical thermodynamics. Derivation of expression for chemical constant and calculation of its value is possible by using of quantum—statistical approach only.

Problem № 3.15

Determine σ and S for helium in CGSM and SI units, if volume of helium equals to 1 L, temperature amd pressure are 400 K and 2 atm respectively.

Solution:

We use expression derived by us (see problem №3.14):

$$\sigma = N\left(\frac{5}{2}\ln T - \ln P + const'\right) \tag{3.64}$$

Let us represent number of particles N by using of mole number:

$$N = nN_A$$

From ideal gas equation:

$$N = \frac{PV}{RT} \times N_A = \frac{2 \times 1 \times 6 \times 10^{23}}{0.082 \times 400} = 0.366 \times 10^{23} \ atoms$$

a) In CGSM units:

$$P = 2\,atm = 2 \times 1.01 \times 10^6 \ \frac{dyne}{cm^2} = 2.02 \times 10^6 \ \frac{dyne}{cm^2}$$

$$const' = 14.74 \ (\text{Table 3.3 in problem № 3.14})$$

$$\sigma = N\left(\frac{5}{2}\ln T - \ln P + const'\right) = 0.366 \times 10^{23}\left(\frac{5}{2}\ln 400 - \ln 2.02 \times 10^6 + 14.74\right) =$$

$$= 0.366 \times 10^{23}(14.98 - 14.52 + 14.74) = 0.366 \times 10^{23} \times 15.2 = 5.56 \times 10^{23}$$

$$S = k\sigma = 1.38 \times 10^{-16} \times 5.56 \times 10^{23} = 7.68 \times 10^7 \ \frac{erg}{K}$$

b) In SI units:

$$P = 2\,atm = 2 \times 1.01 \times 10^5\ Pa = 2.02 \times 10^5\ Pa$$

$$const' = 12.43 \text{ (see Table 3.3 in problem № 3.14)}$$

$$\sigma = 0.366 \times 10^{23} \left(\frac{5}{2} \ln 400 - \ln 2.02 \times 10^5 + 12.43 \right) =$$

$$= 0.366 \times 10^{23} (14.98 - 12.22 + 12.43) = 0.366 \times 10^{23} \times 15.2 = 5.56 \times 10^{23}$$

$$S = k\sigma = 1.38 \times 10^{-23} \times 5.56 \times 10^{23} = 7.68\ J/K$$

Problem № 3.16

The expression for chemical constant (*const'*) was derived by us in problems № 3.12 and № 3.14. The numerical values of *const* and *const'* for helium was calculated also.

1) Convert expressions (3.12) and (3.66) (see Table 3.3) so, that their application for any monoatomic gas will be possible.

2) Calculate the values of chemical constants for argon in CGSM and SI units. Tabulate the obtained results.

Solution:

1)
$$const = \ln \left(\frac{2\pi m k}{h^2} \right)^{3/2} + \frac{5}{2} \tag{3.12}$$

where *m* is molar mass of gas, expressed by *g* or *kg* (according to CGSM or SI units).

Only the values of *m* are different for various gases from quantities in expression (3.12). Let us transform this expression:

$$\ln \left(\frac{2\pi m k}{h^2} \right)^{3/2} + \frac{5}{2} = \ln \left(\frac{2\pi k}{h^2} \right)^{3/2} + \frac{5}{2} + \ln m^{3/2}$$

$$const = \ln\left(\frac{2\pi k}{h^2}\right)^{3/2} + \frac{5}{2} + \ln m^{3/2} \tag{3.67}$$

From (3.66) is obtained analogously:

$$const' = \ln\left(\frac{2\pi k}{h^2}\right)^{3/2} + \frac{5}{2} + \ln k + \ln m^{3/2} \tag{3.68}$$

Let us designate:

$$A = \ln\left(\frac{2\pi k}{h^2}\right)^{3/2} + \frac{5}{2} \tag{a}$$

and

$$B = \ln\left(\frac{2\pi k}{h^2}\right)^{3/2} + \frac{5}{2} + \ln k \tag{b}$$

Let us introduce the numerical values of A and B into (3.67) and (3.68):

CGSM

$$A = \frac{3}{2}\ln\frac{2\pi k}{h^2} + \frac{5}{2} = \frac{3}{2}\ln\frac{6.28 \times 1.38 \times 10^{-16}}{(6.626 \times 10^{-27})^2} + \frac{5}{2} = 131.31$$

$$B = A + \ln k = 131.31 + \ln 1.38 \times 10^{-16} = 131.31 - 36.52 = 94.79$$

$$const = A + \ln m^{3/2} = 131.31 + \frac{3}{2}\ln m \tag{3.69}$$

$$const' = B + \ln m^{3/2} = 94.79 + \frac{3}{2}\ln m \tag{3.70}$$

SI

$$A = \frac{3}{2}\ln\frac{2\pi k}{h^2} + \frac{5}{2} = \frac{3}{2}\ln\frac{6.28 \times 1.38 \times 10^{-23}}{(6.626 \times 10^{-34})^2} + \frac{5}{2} = \frac{3}{2}\ln 1.974 \times 10^{44} = 155.49$$

$$B = \frac{3}{2}\ln\frac{2\pi k}{h^2} + \frac{5}{2} + \ln k = 155.49 + \ln 1.38 \times 10^{-23} = 155.49 - 52.64 = 102.85$$

From these calculations:

$$const = A + \ln m^{3/2} = 155.49 + \frac{3}{2}\ln m \tag{3.71}$$

$$const' = B + \ln m^{3/2} = 102.85 + \frac{3}{2}\ln m \tag{3.72}$$

2) Let us calculate the values of chemical constants for argon in CGSM and SI units. Relative molar mass of argon is equal to 40.

CGSM

$$M_{Ar} = 40 \times 1.66 \times 10^{-24} = 6.64 \times 10^{-23}\ g$$

According to (3.69),

$$const = 131.31 + \frac{3}{2}\ln 6.64 \times 10^{-23} = 131.31 - 76.6 = 54.71$$

From (3.70) we obtain:

$$const' = 94.79 + \frac{3}{2}\ln 6.64 \times 10^{-23} = 94.79 - 76.6 = 18.19$$

SI

$$M_{Ar} = 40 \times 1.66 \times 10^{-27} = 6.64 \times 10^{-26}\ kg$$

According expression (3.71),

$$const = 155.49 + \frac{3}{2}\ln 6.64 \times 10^{-26} = 155.49 - 86.96 = 68.53$$

It follows from (3.72):

$$const' = 102.85 + \frac{3}{2}\ln 6.64 \times 10^{-26} = 102.85 - 86.96 = 15.89$$

Let us tabulate the results obtained in problems №№ 3.12, 3.14 and 3.16.

Table 3.4. Expressions of chemical constants and their numerical values in SI and CGSM units for monoatomic ideal gas

Expression for chemical constant(a)	Expression for chemical constant (b)		Numerical value of chemical constant			
	CGSM	SI	CGSM		SI	
			He	Ar	He	Ar
$const = \ln\left(\frac{2\pi k}{h^2}\right)^{3/2} +$ $+\frac{5}{2} + \ln m^{3/2}$ (3.67)	$const = 131.31 +$ $+\frac{3}{2}\ln m$ (3.69)	$const = 155.49 +$ $+\frac{3}{2}\ln m$ (3.71)	51.26	54.71	65.07	68.53
$const' = \ln\left(\frac{2\pi k}{h^2}\right)^{3/2} +$ $+\frac{5}{2} + \ln k + \ln m^{3/2}$ (3.68)	$const' = 94.79 +$ $+\frac{3}{2}\ln m$ (3.70)	$const' = 102.85 +$ $+\frac{3}{2}\ln m$ (3.72)	14.74	18.19	12.43	15.89

Problem № 3.17

Determine entropy value of 1 L of helium under the following conditions:
1) 5 K, 150 atm; 2) 10 K, 1000 atm; 3) 100 K, 1 atm; 4) 300 K, 1 atm;

5) 500 K, 100 atm. Explain the obtained results.

Solution:

Let use Sacure—Tetrode's equation:

$$S = kN\left(\frac{5}{2}\ln T - \ln P + const'\right) \tag{3.65}$$

If take into account, that $kN=Rn,$ then it is obtained from (3.65):

$$S = Rn\left(\frac{5}{2}\ln T - \ln P + const'\right) \tag{3.73}$$

In SI units $R = 8.31\ J/mol \times K,\ const' = 12.43$ (see Table 3.3 in problem № 3.14).

1) $T = 5K,\ P = 150\,atm = 150 \times 1.01 \times 10^5 = 151.5 \times 10^5\ Pa$

$$n = \frac{PV}{RT} = \frac{150 \times 1}{0.082 \times 5} = 365.85\ mole$$

$$S = Rn\left(\frac{5}{2}\ln T - \ln P + const'\right) = 8.31 \times 366 \times \left(\frac{5}{2}\ln 5 - \ln 151.5 \times 10^5 + const'\right) =$$

$$= 3040.25(4.02 - 16.53 + 12.43) = -243.22\ J/K$$

2) $T = 10K,\ P = 1000\,atm = 1.01 \times 10^8\ Pa$

$$n = \frac{PV}{RT} = \frac{1000 \times 1}{0.082 \times 10} = 1219.51\ moles$$

$$S = 1219.51 \times 8.31\left(\frac{5}{2}\ln 10 - \ln 1.01 \times 10^8 + 12.43\right) = 10134(5.76 - 18.43 + 12.43) =$$

$$= -2432.16\ J/K$$

3) $T = 100\,K,\ P = 1\,atm = 1.01 \times 10^5\ Pa$

$$n = \frac{PV}{RT} = \frac{1 \times 1}{0.082 \times 100} = 0.12 \ mole$$

$$S = 0.12 \times 8.31\left(\frac{5}{2}\ln 100 - \ln 1.01 \times 10^5 + 12.43\right) = 1.01(11.51 - 11.52 + 12.43)\Bigg) =$$

$$= 12.54 \ J/K$$

   4) $T = 300 \ K, \ P = 1 \, atm = 1.01 \times 10^5 \ Pa$

$$n = \frac{PV}{RT} = \frac{1 \times 1}{0.082 \times 300} = 0.041 \ mole$$

$$S = 0.041 \times 8.31\left(\frac{5}{2}\ln 300 - \ln 1.01 \times 10^5 + 12.43\right) = 0.338(14.26 - 11.52 + 12.43) =$$

$$= 5.13 \ J/K$$

   5) $T = 500 \ K, \ P = 100 \, atm = 1.01 \times 10^7 \ Pa$

$$n = \frac{PV}{RT} = \frac{100 \times 1}{0.082 \times 500} = 2.44 \ moles$$

$$S = 2.44 \times 8.31\left(\frac{5}{2}\ln 500 - \ln 1.01 \times 10^7 + 12.43\right) = 20.27(15.54 - 16.13 + 12.43) =$$

$$= 240 \ J/K$$

Let us tabulate the obtained results (Table 3.5).

It is seen from Table 3.5, that entropy has a negative value at low temperatures (5÷10 K) and high pressures (150÷1000 atm); but according to Planck's exclusion ($S \geq 0$), entropy can not accept a negative value. Such luck of correspondence lies in the fact that at low temperature and high pressure helium exists in quantum regime and does not represent an ideal gas (see Table 3.2 in problem № 3.11). Sacure—Tetrode's equation may be used for an ideal gas only.

As regards comparatively high temperatures ($T \geq 100$ K), gas already exists in the classical conditions (even under high pressure) and entropy, calculated by Sacure—Tetrode's equation has a positive value.

Table 3.5.

$$V_{He} = 1 \text{ L}$$

T K	P, atm	S, J/K	Regime	Gas
5	150	-243	Quantum	Nonideal
10	1000	-2432	Quantum	Nonideal
100	1	12	Classical	Ideal
300	1	5	Classical	Ideal
500	100	240	Classical	Ideal

Problem №3.18

The ideal gas is placed in the container at the conditions of P pressure and T temperature. The container is divided into two equal parts by the barrier. Both parts contain the same number of moles (n) and take the same volume (V). When the barrier is taken away, then the "combined" system is obtained, which contains $2n$ mole of compound and takes the volume $2V$.

Use the expression $S\ (V,T) = nS^0(T) + nR\ lnV$ and establish whether the principle of entropy additivity: $S_{"comb."} = S_1 + S_2$ is executed for this system.

Solution:

$$S(V,T) = nS^0(T) + nR \ln V \qquad (3.74)$$

where $nS^0\ (T)$ is the integration constant, which depends on the amount of component and temperature.

According to the expression (3.74) and the condition of problem, entropies of first and second systems before taking the barrier away are equal:

$$S_1 = S_2 = nS^0(T) + nR \ln V$$

Entropy sum of these two systems is:

$$S_1 + S_2 = 2nS^0(T) + 2nR\ln V$$

Entropy of the "combined" system obtained after taking the barrier away is equal:

$$S_{comb.} = 2nS^0(T) + 2nR\ln 2V$$

In accordance with entropy additivity principle, S_1+S_2 must be equal to $S_{\text{"comb."}}$. Let see if this is so:

$$S_{comb.} - (S_1 + S_2) = 2nS^0(T) + 2nR\ln 2V - 2nS^0(T) - 2nR\ln V =$$
$$= 2nR\ln 2V - 2nR\ln V = 2nR\ln 2 \neq 0$$
$$\text{i.e. } S_{comb.} \neq S_1 + S_2$$

Thus, either entropy additivity principle is incorrect, or the expression (3.74) is erroneous. Deducing of other expression for $S(V,T)$ by means of classical thermodynamics is impossible. On the other hand, the negation of entropy additivity principle deprives off the foundation of thermodynamics. This outstanding problem is known as Gibbs's paradox, because Willard Gibbs has first considered this contemplative experiment. It should be noted, that Ludwig Boltzmann and Willard Gibbs have corrected expression (3.74) at their discretion and instead of it used the formula:

$$S(V,T) = nS^0(T) + nR\ln\frac{V}{n} \tag{3.75}$$

In this case entropy additivity principle is not disturbed, however deducing of expression (3.75) from some other equation is not succeeded.

The avoidance of Gibbs's paradox became possible later, by using of statistical thermodynamics and quantum mechanics.

Problem № 3.19

Use expression $S\ (P,T\ )= nS^0\ (T\ )\ -\ nR\ lnP$ and establish, if entropy additivity principle is fulfilled for systems considered in problem № 3.18.

Solution:

Entropy of individual systems before taking the barrier away is equal:

$$S_1 = S_2 = nS^0(T) - nR \ln P$$

Sum of entropy of the both systems is:

$$S_1 + S_2 = 2nS^0(T) - 2nR \ln P$$

Entropy of "combined" system after taking the barrier away is equal:

$$S_{comb.} = 2nS^0(T) - 2nR \ln P$$

$$S_{comb.} - (S_1 + S_2) = 2nS^0(T) - 2nR \ln P - 2nS^0(T) + 2nR \ln P = 0$$

$$i.e.\ S_{comb.} = S_1 + S_2.$$

Thus, in contrast to $S\ (V,T)$, entropy additivity principle is not disturbed by using expression for $S\ (P,T)$.

Problem № 3.20

Use Sacure—Tetrode's equation:

$$S = Rn\left(\frac{3}{2}\ln T - \ln\frac{N}{V} + const\right) \tag{3.13}$$

and estimate if entropy additivity principle is executed for the systems considered in problem № 3.18.

Solution:

According to (3.13) entropy of each individual system before taking the barrier away is equal:

$$S_1 = S_2 = Rn\left(\frac{3}{2}\ln T - \ln\frac{N}{V} + const\right)$$

The sum of entropy of the both system is:

$$S_1 + S_2 = 2Rn\left(\frac{3}{2}\ln T - \ln\frac{N}{V} + const\right)$$

Entropy of the "combined" system, obtained as a result of taking the barrier away is equal:

$$S_{comb.} = 2Rn\left(\frac{3}{2}\ln T - \ln\frac{2N}{2V} + const\right)$$

$$S_{comb.} = S_1 + S_2$$

Thus, the condition of entropy additivity is realized by using of Sacure—Tetrode's equation.

Problem № 3.21

Function of which quantities represents integration constant $nS^0(T)$ in expression:

$$S = nS^0(T) + nR\ln V \qquad (3.74)$$

Calculate the values of integration constant and entropy for 1 L of helium under conditions 300 K and 1 atm.

Solution:

It was considered early, that $nS^0(T)$ is a function of amount of substance and temperature only. It was revealed later, that $nS^0(T)$ is a function of some other quantity also, neglecting of which results in Gibbs's paradox (see problem № 3.18). In order to elucidate, on which quantities is depended $nS^0(T)$, let us refer to Sacure—Tetrode's equation:

$$S = Rn\left(\frac{3}{2}\ln T - \ln\frac{N}{V} + const\right) \tag{3.13}$$

where N is number of particles and

$$const = \ln\left(\frac{2\pi mk}{h^2}\right)^{3/2} + \frac{5}{2} \tag{3.12}$$

m represents molar mass, h and k are Planck's and Boltzmann's constants respectively.

Let us equalize the right parts of expressions (3.74) and (3.13):

$$nS^0(T) + Rn\ln V = Rn\frac{3}{2}\ln T + Rn\ln V - Rn\ln N + Rn \times const$$

It follows that

$$nS^0(T) = Rn\left(\frac{3}{2}\ln T - \ln N + const\right) \tag{3.76}$$

or

$$nS^0(T) = kN\left(\frac{3}{2}\ln T - \ln N + const\right) \tag{3.77}$$

Thus, $nS^0(T)$ represents a function not only of $\ln T$ and N, but also of $\ln m$ and m.

Let us determine the value of $nS^0(T)$ for 1 L of helium under the conditions 300 K and 1 *atm* pressure (In SI units).

$$n = \frac{PV}{RT} = \frac{1 \times 1}{0.082 \times 300} = 0.04 \, mole; \quad N = 0.04 \times 6 \times 10^{23} = 2.4 \times 10^{22} \, atoms$$

$const = 65.07$ (see Table 3.3 in problem № 3.14).

Let us introduce these values into expression (3.76):

$$nS^0(T) = nR\left(\frac{3}{2}\ln T - \ln N + const\right) = 0.04 \times 8.31 \times$$

$$\times \left(\frac{3}{2}\ln 300 - \ln 2.4 \times 10^{22} + 65.07\right) = 0.04 \times 8.31 \times 22.09 = 7.34 \; J/K$$

Entropy value is equal to:

$$S(V, T) = nS^0(T) + nR \ln V = 7.34 + 0.04 \times 8.31 \times \ln 1 \times 10^{-3} =$$

$$= 7.34 - 2.30 = 5.04 \; J/K$$

Problem № 3.22

Check–up, occurs or not Gibbs's paradox, if integration constant in the (3.74) expression: $S(V,T) = nS^0(T) + nR \ln V$ will be substituted by its equivalent expression:

$$nS^0(T) = Rn(\frac{3}{2}\ln T - \ln N + const) \qquad (3.76)$$

Solution:

Let designate:

$$A = \frac{3}{2}\ln T + const,$$

Then it is obtained from (3.76):

$$nS^0 T = nR(A - \ln N) \tag{a}$$

Let enter (a) into (3.74):

$$S = Rn(A - \ln N) + Rn \ln V = Rn(A - \ln \frac{N}{V}) \tag{b}$$

According to (b), entropies of first and second systems untill taking away a barrier are equal to:

$$S_1 = S_2 = Rn\left(A - \ln \frac{N}{V}\right).$$

The sum of entropies of these systems is:

$$S_1 + S_2 = 2Rn\left(A - \ln \frac{N}{V}\right)$$

The entropy of the combined system, obtained after taking away the barrier, is depicted following:

$$S_{comb.syst.} = 2Rn\left(A - \ln \frac{2N}{2V}\right) = 2Rn\left(A - \ln \frac{N}{V}\right)$$

$$S_{comb.syst.} = S_1 + S_2,$$

i.e. the additivity of entropy is fulfilled.

Thus, it is possible to avoid Gibbs's paradox, taking into account, that the constant of integration represents not only the function of $\ln T$ and N, but also of $\ln N$.

The question is arisen: why the existence of $\ln N$ in the expression of entropy has such a great significance? After all, this expression includes N and the value of entropy should be depended on N much stronger, than on $\ln N$.

The fact is that, the entering of $\ln N$ in entropy expression causes the transformation of $\ln V$ into $(-\ln C)$: $\ln V - \ln N = -\ln \dfrac{N}{V} = -\ln C$. The difference between V and $C = \dfrac{N}{V}$ is considerable. It is obvious, that the use of the volume instead of the concentration may mislead us. Really, the volume of the "combined" system is two times that of the separate systems, but the concentration in all three systems is the same. Therefore the principle of additivity of entropy is disturbed, if it is taken into account the volume only. But taking into consideration that $C = \dfrac{N}{V}$, the abovementioned principle is fulfilled. (Gibbs's paradox does not take place when pressure is used instead of volume, because pressure represents the value, proportional to the concentration: $P=CRT$). (See problem № 3.19).

Problem № 3.23

1 mole of He is placed in the container at pressure 2 atm and ambient temperature. The system expands adiabatic:

1) Reversible till pressure 1 atm;

2) Irreversible opposite to pressure 1 atm.

Calculate entropy changes of system, surroundings and "universe" in both cases.

(Use the results of problem № 1.16).

Solution:

$$V_1 = \frac{nRT_1}{P_1} = \frac{1 \times 0.082 \times 298}{2} = 12.22 \ L$$

1) After reversible adiabatic expansion $T_2 = 225,9\ K, \quad V_2 = 18.52\ L$ (see Table 1.3 in problem № 1.16).

Introduce these data in expression (3.16):

$$\Delta S_{syst.} = nR\left(\frac{3}{2}\ln\frac{T_2}{T_1} + \ln\frac{V_2}{V_1}\right) = 1\times8,31\left(\frac{3}{2}\ln\frac{225,9}{298} + \ln\frac{18.52}{12.22}\right) =$$

$$= 1\times8,31(-0,416 + 0,416) = 0$$

According to expressions (3.44) and (3.41):

$$\Delta S_{surr.} = \frac{q_{surr.}}{T} = \frac{0}{T} = 0$$

$$\Delta S_{"univ."} = \Delta S_{syst.} + \Delta S_{surr.} = 0$$

2) In accordance with Clausius's inequality,

$$\Delta S_{syst.}^{irrev.} > \frac{q^{irrev.}}{T}_{syst.} \tag{3.22}$$

In adiabatic process $q_{rev.} = q_{irrev.} = 0$ i.e.

$$\Delta S_{syst.}^{irrev.} > 0$$

As a result of irreversible adiabatic expansion the characterizing parameters of system take following values: $T_2 = 238,5$ K and $V_2 = 19.56\ L$ (see Table 1.3 in problem № 1.16).

According to (3.16)

$$\Delta S_{syst.} = nR\left(\frac{3}{2}\ln\frac{T_2}{T_1} + \ln\frac{V_2}{V_1}\right) = 1\times8,31\left(\frac{3}{2}\ln\frac{238,5}{298} + \ln\frac{19.56}{12.22}\right) =$$

$$= 8,31(-0,334 + 0,470) = 1.13 \ J/K$$

Entropy change of surroundings is stipulated by heat exchange with system only. In adiabatic process $q_{surr.} = q_{syst.} = 0$. Therefore

$$\Delta S_{surr.} = \frac{q_{surr.}}{T} = 0$$

$$\Delta S_{"univ."} = \Delta S_{syst.} + \Delta S_{surr.} = 1.13 + 0 = 1.13 \ J/K$$

Remark: The system transfers from the same initial conditions into the different final states due to reversible and irreversible adiabatic expansion. The expansion in the irreversible process is completed at higher values of T and V, than in reversible one. In the irreversible process the system (in more or less extent) remains without supervision and attempts to possess $S_{fin.}$ and ΔS as much as possible. If remember, that $S = f(T,V,N)$, then it is clear why the values of T_2 and V_2 are higher at irreversible adiabatic expansion, than at reversible one and why $\Delta S_{irrev.} > \Delta S_{rev.}$ (See also problem № 1.16).

Problem № 3.24
1 mole of He is placed under pressure 2 atm at ambient temperature. It expands reversibly and isothermally till pressure 1 atm. Determine:
   1) work fulfilled by system;
   2) the amount of heat, exchanged between system and surroundings;
   3) entropy change of system, surroundings and "universe".

Solution:
According to (1.36), work fulfilled by system in reversible isothermal process is equal:

$$A_{syst.}^{rev.} = nRT \ln \frac{P_2}{P_1} = 1 \times 8,31 \times 298 \ln \frac{1}{2} = -1716.5 \ J$$

The internal energy of ideal gas does not change in isothermal process:

$$\Delta U = q + A = 0$$

From this equality

$$q_{syst.}^{rev.} = -A_{syst.}^{rev.} = 1716.5 \ J$$

According to expression (3.42),

$$q_{surr.}^{rev.} = -q_{syst.}^{rev.} = -1716.5 \ J$$

By expression (3.33) at reversible isothermal expansion

$$\Delta S_{syst.}^{rev.} = nR \ln \frac{P_1}{P_2} = 1 \times 8,31 \ln \frac{2}{1} = 5.76 \ J/K$$

In reversible process

$$\Delta S_{surr.}^{rev.} = -\Delta S_{syst.}^{rev.} = -5.76 \ J/K$$

$$\Delta S_{"univ."}^{rev.} = \Delta S_{syst.}^{rev.} + \Delta S_{surr.}^{rev.} = 5.76 - 5.76 = 0$$

Problem № 3.25

1 mole of He is placed at 298 K and 2 atm pressure. It expands isothermal and irreversible opposite to 1 atm pressure. Determine:
1) The amount of work fulfilled by system;
2) The values of q of system and surroundings;
3) Entropy change of system, surroundings and "universe".

Solution:
1) The work fulfilled by system at irreversible expansion proceeding opposite to constant external pressure is expressed so:

$$A_{syst.}^{irrev.} = -P_{ex}\Delta V \qquad (1.8)$$

From the ideal gas equation:

$$V_1 = \frac{nRT}{P_1} = \frac{1 \times 0.082 \times 298}{2} = 12.22 \ L$$

$$V_2 = \frac{nRT}{P_2} = \frac{1 \times 0.082 \times 298}{1} = 24.44 \ L$$

$$\Delta V = V_2 - V_1 = 24.44 - 12.22 = 12.22 \ L$$

$$A_{syst.}^{irrev.} = -P_{ex}\Delta V = -1 \times 12.22 \ L = -12.22 \ L \times atm = -12.22 \times 101.34 \ J =$$

$$= -1238.38 \ J$$

2) The change of internal energy of system in isothermal conditions is:

$$\Delta U = q + A = 0$$

hence

$$q_{syst.}^{irrev.} = -A_{syst.}^{irrev.} = 1238.38 \ J$$

According to expression (3.42), in both reversible and irreversible processes:

$$q_{surr.} = -q_{syst.} = -1238.38 \ J$$

3) From the comparison of problems № 3.24 and № 3.25 follows, that the initial and final states are the same at the reversible and irreversible isothermal expansions. Since entropy is state function, therefore

$$\Delta S_{syst.}^{irrev.} = \Delta S_{syst.}^{rev.} = 5.76 \ J / K$$

According to expressions (3.44b) and (3.41),

$$\Delta S_{surr.}^{irrev.} = \left(\frac{q}{T}\right)_{surr.}^{irrev.} = -\frac{1238.38}{298} = -4.15 \, J/K$$

$$\Delta S_{"univ."}^{irrev.} = \Delta S_{syst.}^{irrev.} + \Delta S_{surr.}^{irrev.} = 5.76 - 4.15 = 1.61 \, J/K$$

Remark: For **system** the values of q are different in reversible and irreversible processes, but ΔS are the same. According Clausius's expressions (3.21) and (3.22),

$$\Delta S_{syst.}^{rev.} = \left(\frac{q}{T}\right)_{syst.}^{rev.} = \frac{1716.5}{298} = 5.76 \, J/K$$

$$\Delta S_{syst.}^{irrev.} > \left(\frac{q}{T}\right)_{syst.}^{irrev.} = \frac{1238.38}{298} = 4.15 \, J/K$$

Since entropy of system is state function, therefore

$$\Delta S_{syst.}^{irrev.} = \Delta S_{syst.}^{rev.} = 5.76 \, J/K$$

For **surroundings** the q values are different in reversible and irreversible processes, moreover, the ΔS values are also distinguished. This is caused by equality (3.44), taking place for surroundings in both reversible and irreversible processes:

$$\Delta S_{surr.} = \frac{q_{surr.}}{T} \qquad (3.44)$$

$$\Delta S_{surr.}^{rev.} = \left(\frac{q}{T}\right)_{surr.}^{rev.} = -\frac{1716.5}{298} = -5.76 \, J/K$$

$$\Delta S_{surr.}^{irrev.} = \left(\frac{q}{T}\right)_{surr.}^{irrev.} = -\frac{1238.38}{298} = -4.15 \, J/K$$

$$\Delta S_{surr.}^{rev.} \neq \Delta S_{surr.}^{irrev.}$$

Such difference between entropy changes of system and surroundings is caused with assumption that entropy change of surroundings is stipulated by heat exchange with system only, but entropy change of system besides of heat exchange is determined by other factors also (e.g. changes of volume and (or) pressure).

Problem № 3.26

1 mole of He is placed in adiabatic container at one side of the barrier under the pressure 2 atm. On the other side is the emptiness. The volume of gas is doubled after taking the barrier away. The expansion begins at 298 K.

1) Calculate *A, q,* ΔU and ΔT of this process.

2) Determine entropy changes of system, surroundings and "universe".

Solution:

1) The expansion occurs in a vacuum under constant external pressure $P_{ex} = 0$. Therefore the process is irreversible. Following expression (1.8),

$$A_{irrev.} = -P_{ex}\Delta V = 0$$

According to conditions of problem, the process is adiabatic *i.e.* $q = 0$. The change of internal energy of system is equal:

$$\Delta U = q + A = 0$$

The internal energy of ideal gas is temperature function only: $U = f(T)$. If internal energy does not changes at the proceeding of process ($\Delta U = 0$) then temperature is also invariable: $\Delta T = 0$.

Thus, the expansion of gas in a vacuum is adiabatic and isothermal process simultaneously. During the process work is not fulfilled, the heat exchange does not take place, the temperature and internal energy remain invariable. The only, what changes, is volume (and hence pressure). The increase of volume just represents the moving force of this spontaneous, maximal irreversible process, because during this process the occupation of new coordinates and hence of new energy levels i.e. the transfer of system into the state with much greater probability takes place. The greater probability, in its turn, corresponds to the higher value of entropy:

$$S = k \ln W \tag{3.1}$$

where W represents the thermodynamic probability of system's state.

2) Entropy change at maximal irreversible isothermal expansion should be the same as in the reversible and less irreversible isothermal expansion, discussed in problems №№ 3.24 and 3.25 (because the initial and final states are the same in these processes).

According to expression (3.33), at isothermal expansion

$$\Delta S_{syst.} = nR \ln \frac{V_2}{V_1} = 1 \times 8,31 \ln \frac{2}{1} = 5.76 \ J/K$$

As is already mentioned, the expansion in a vacuum represents not only isothermal but also adiabatic process: $q_{syst.} = q_{surr.} = 0$. Therefore

$$\Delta S_{surr.} = \frac{q_{surr.}}{T} = 0$$

$$\Delta S_{"univ."} = \Delta S_{syst.} + \Delta S_{surr.} = 5.76 + 0 = 5.76 \ J/K$$

Problem № 3.27

Find the values of "production" of entropy ($\Delta_i S$) in the irreversible processes discussed in problems № 3.23, № 3.25 and № 3.26.

Solution:
According to the expression (3.49):

$$\Delta S_{syst.} = \Delta_e S + \Delta_i S$$

where $\Delta_e S$ is entropy change of the system, caused by the heat exchange with surroundings and $\Delta_i S$ – entropy change of the system, induced by the irreversibility of the process itself (by the "production" of entropy).

It is accepted, that entropy change of surroundings is stipulated only by the heat exchange with system and not by some other factors:

$$\Delta S_{surr.} = \frac{q_{surr.}}{T} = -\frac{q_{syst.}}{T} \qquad (3.44)$$

Consequently

$$\Delta S_{surr.} = -\Delta_e S \qquad (3.78)$$

Let us enter expressions (3.49) and (3.78) into the (3.41). Then it will be obtained:

$$\Delta S"_{univ."} = \Delta S_{syst.} + \Delta S_{surr.} = \Delta_e S + \Delta_i S - \Delta_e S = \Delta_i S$$

Thus, entropy change of "universe" is equal to "production" of entropy in the system:

$$\Delta S"_{univ."} = \Delta_i S \qquad (3.79)$$

Entropy is not produced in the reversible process:

$$\Delta_i S = 0 \qquad \text{and} \qquad \Delta S"_{univ."} = 0$$

Entropy is induced in the irreversible spontaneous process:

$$\Delta_i S > 0 \qquad \text{and} \qquad \Delta S"_{univ."} > 0$$

Expression (3.79) is correct for all – reversible and spontaneous irreversible – processes. But in the specific cases $\Delta_i S$ may be determined otherwise also.

1) Irreversible adiabatic expansion (problem № 3.23). As we have seen,

$$\Delta_e S = \frac{q}{T}$$

In the adiabatic process $q = 0$ and $\Delta_e S = 0$. According to (3.49),

$$\Delta_i S = \Delta S_{syst.} - \Delta_e S = \Delta S_{syst.} = 1.13\, J\,/\,K$$

Thus, entropy change of the system at irreversible adiabatic expansion is wholly stipulated by the "production" of entropy; and the last – by the spontaneity of the process.

2) Irreversible isothermal expansion (problem № 3.25). According to expression (3.49), for the system:

$$\Delta_i S_{irrev.} = \Delta S_{irrev.} - \Delta_e S_{irrev.} = \Delta S_{irrev.} - \frac{q_{irrev.}}{T}$$

If the initial and final conditions of reversible and irreversible processes are alike, then $\Delta S_{rev.} = \Delta S_{irrev.}$ Therefore,

$$\Delta_i S_{irrev.} = \Delta S_{rev.} - \frac{q_{irrev.}}{T}$$

In problems № 3.24 and № 3.25 we have calculated that $\Delta S_{rev.} = 5.76 J/K$ and $\frac{q_{irrev.}}{T} = 4.15\, J\,/\,K$ i.e.

$$\Delta_i S_{irrev.} = 5.76 - 4.15 = 1.61\, J\,/\,K$$

3) Expansion in a vacuum (problem № 3.26). The process is adiabatic:

$$q = 0 \qquad \text{and} \qquad \Delta_e S = \frac{q}{T} = 0$$

According to (3.49),

$$\Delta_i S = \Delta S_{syst.} - \Delta_e S = \Delta S_{syst.} = 5.76 \ J \, / \, K$$

Let us tabulate the data according to the results of problems №№ 3.23 ÷ 3.27.

Table 3.6. The characteristic values of the expansions of different type.

(1 mol He, T_1 =298 K, V_1 = 12.22 L, P_1 = 2 atm, P_2 = 1 atm)

Expansion	T_2 K	V_2 L	$q_{syst.}$ kJ	$q_{surr.}$ kJ	$A_{syst.}$ kJ	ΔS, J/K System	Surroundings	"Universe"	$\Delta_i S$, J/K (Production of entropy)
Adiabatic									
Reversible	225.9*	18.5*	0	0	–0.9*	0	0	0	0
Irreversible	238.5*	19.6*	0	0	–0.7*	1.13	0	1.13	1.13
Isothermal									
Reversible	298	24.44	1.7	–1.7	–1.7	5.76	–5.76	0	0
Irreversible (P_{ex} = const) $P_{ex} \neq 0$	298	24.44	1.2	–1.2	–1.2	5.76	–4.15	1.61	1.61
P_{ex}=0 (expansion in a vacuum)	298	24.44	0	0	0	5.76	0	5.76	5.76

*data are taken from problem № 1.16

As can be seen from the Table 3.6:

- The final states of the system are distinguished in reversible and irreversible processes after adiabatic expansion. Therefore, $\Delta S_{rev.} \neq \Delta S_{irrev.}$ (See also problem № 1.16).
- The final states of the system are the same in the reversible ($P_{ex} \neq$ const), irreversible ($P_{ex} =$ const $\neq 0$) and extremely irreversible ($P_{ex} = 0$) processes of isothermal expansion. As a result, $\Delta S_{syst.}$ are equal in all three cases, but $\Delta S_{surr.}$ increases with the enhancement of degree of irreversibility.
- The production of entropy is equal to the change of entropy of "universe" (in both reversible and irreversible processes). $\Delta_i S$ increases with the spontaneity of the process. In the extremely irreversible process (expansion in a vacuum, $A = 0$) increasing of entropy of the system is completely stipulated by the irreversible character of the process itself, i.e. by the production of entropy: $\Delta S"_{univ."} = \Delta S_{syst.} = \Delta_i S$.
- The system fulfills more work in the reversible process than in the irreversible one. Besides, work of the adiabatic process is less than the work of isothermal process (see also problem № 1.11 and Table 1.3).

Problem № 3.28

Two different gases are present on the left and right sides of barrier, by which the container is divided. Number of moles and volume of the first gas are n_1 and V_1, but mole number and volume of the second gas are equal to n_2 and V_2 respectively. The isothermal mixing of gases occurs by taking the barrier away.

Derive expression of entropy change in this process.

Solution:

Diffusion of gases represents an irreversible spontaneous process. Nevertheless, entropy change of this process may be determined by means of corresponding (proceeded in the same initial and final conditions) reversible process. The last may be represented as expansion of each gas from volume V_1 and V_2 until ($V_1 + V_2$) one.

At isothermal expansion

$$\Delta S = Rn \ln \frac{V_{fin.}}{V_{init.}} \qquad (3.33)$$

Entropy change in case of expansion of the first gas is:

$$\Delta S_1 = Rn_1 \ln \frac{V_1 + V_2}{V_1}$$

Entropy of expansion of second gas is:

$$\Delta S_2 = Rn_2 \ln \frac{V_1 + V_2}{V_2}$$

Following the principle of entropy additivity,

$$\Delta S_{mix.} = \Delta S_1 + \Delta S_2 = Rn_1 \ln \frac{V_1 + V_2}{V_1} + Rn_2 \ln \frac{V_1 + V_2}{V_2}$$

If take into consideration, that

$$\frac{V_1}{V_1 + V_2} = \frac{n_1}{n_1 + n_2} = \kappa_1 \qquad \text{and} \qquad \frac{V_2}{V_1 + V_2} = \frac{n_2}{n_1 + n_2} = \kappa_2,$$

where κ_1 and κ_2 are the molar fractions of first and second gases in the mixture, then

$$\Delta S_{mix.} = Rn_1 \ln \frac{1}{\kappa_1} + Rn_2 \ln \frac{1}{\kappa_2} = -R(n_1 \ln \kappa_1 + n_2 \ln \kappa_2)$$

In case of *i–th* component

$$\Delta S_{mix.} = -R \sum n_i \ln \kappa_i \qquad (3.35)$$

Since $\kappa_i < 1$, therefore $\ln \kappa_i < 0$ and $\Delta S_{mix.} > 0$.

Thus, mixing of ideal gases represents a spontaneous process, which is characterized by entropy increase at the either ratios of constituents.

Problem 3.29

Calculate entropy change by formation of a) 10 moles and b) 1 mole of air. Compare these values with one another.

Solution:

Formation of air means the mixing of 80% N_2 and 20% O_2. Therefore $\Delta S_{form.} = \Delta S_{mix.}$

a) There are 8 moles of N_2 and 2 moles of O_2 in the 10 mole of air.

$$\kappa_{N_2} = \frac{8}{2+8} = 0.8 \qquad \text{and} \qquad \kappa_{O_2} = \frac{2}{2+8} = 0.2$$

According to expression (3.35):

b) 1 mole of air contains 0.8 mole of N_2 and 0.2 mole of O_2.

$$\kappa_{N_2} = \frac{0.8}{0.2+0.8} = 0.8 \qquad \text{and} \qquad \kappa_{O_2} = \frac{0.2}{0.2+0.8} = 0.2$$

In case, when $\sum n_i = 1$, expression (3.36) may be used:

$$\Delta S_{mix.} = -R\sum \kappa_i \ln \kappa_i = -8.31(0.8\ln 0.8 + 0.2\ln 0.2) = 4.16 \ J/K$$

Let us compare the entropy values of formation of 1 and 10 *moles* of air:

$$\Delta S\,(10\ moles) = 41.6\ J/K \quad \text{and} \quad \Delta S\,(1\ mole) = 4.16\ J/K.$$

$$41.6 = 10 \times 4.16$$

i.e..$\Delta S\,(n) = n\Delta S$, which means that entropy is an additive quantity.

Problem № 3.30

There is an ideal gas at pressure P and temperature T in the container, divided by the barrier. The mole number of gas at one side of barrier is n_1, the volume is equal to V_1. On the other side of barrier amount of moles is n_2 and the volume is V_2.

Determine ΔS in the case when the barrier is taken away.

Solution:

As a result of taking away the barrier, the different parts of the same gas are mixed. In order to determine the change of entropy in this process, the expression (3.35) is used:

$$\Delta S_{mix.} = -R(n_1 \ln \kappa_1 + n_2 \ln \kappa_2)$$

If we take into account, that the same gas is at the both sides of the barrier, then in the mixture:

$$\kappa_1 = \kappa_2 = \frac{n_1 + n_2}{n_1 + n_2} = 1$$

and

$$\Delta S_{mix.} = -R(n_1 \ln 1 + n_2 \ln 1) = 0$$

Remark: Gibbs's paradox does not take place in this case because the concentration (κ) is included in the expression (3.35) (in more details see problem № 3.22).

Problem № 3.31

Entropy of solid solutions is not equaled to zero even at the temperature 0 K. The solid solution possesses so-called "residual entropy". What is the value of "residual entropy" of 1 mole binary solid solution, if entropy of the constituents of solution is equal to zero at 0 K and their molar fractions are the same?

Solution:

Entropy change by the formation of a solution is equal:

$$\Delta S_{mix.} = S_{solut.} - S_1 - S_2,$$

where S_1 and S_2 – entropies of components before the mixing.

According to the condition of sum, $S_1 = S_2 = 0$ at 0 K. Then

$$\Delta S_{mix.} = S_{solut.}$$

Thus, "residual entropy" of solution at zero K is equal to entropy of mixing.

Following the condition of sum, $\kappa_1 = \kappa_2 = 0,5$. Let enter this value in expression (3.36):

$$\Delta S_{mix.} = -R\sum \kappa_i \ln \kappa_i = -R(0,5\ln 0,5 + 0,5\ln 0,5) = -8,31 \cdot \ln 0,5 = 5,76 \ \ J/K$$

Eventually, when $\kappa_1 = \kappa_2 = 0,5$, then the residual entropy of the binary solid solution is equal to $5,76 \ J/K$ at absolute zero.

Problem № 3.32

It is established, that molar entropy of CO at absolute zero is $4.6 \ J / K \times mol$. Does this fact contradict the third law of thermodynamics?

Solution:

The third law of thermodynamics : $S_0 = 0$ is correct for ideal crystals only. Entropy of nonideal crystal is nonzero, which is stipulated by certain irregularity of particles' arrangement in the lattice points. It may be also conditioned by existence of vacancies and other defects.

Presumably, molecules in crystalline lattice of solid CO arrange in disorder: CO, OC, OC, CO, OC, CO, CO.... Due to such arrangment it may be supposed, that crystal represents the mixture of CO and OC. Its "residual" entropy is equal to the entropy of mixing (see problem № 3.31):

$$S_{CO} = \Delta S_{mix.} = -R(\kappa_{CO} \ln \kappa_{CO} + \kappa_{OC} \ln \kappa_{OC})$$

If assume, that arrangment of molecules has quite random character, then $\kappa_{CO} = \kappa_{OC} = 0.5$ and

$$S_{CO} = -R(0.5 \ln 0.5 + 0.5 \ln 0.5) = -8.31 \times (-0.693) = 5.76 \ J/K$$

But according to condition of problem, $S_{CO} = 4.6 \ J/K$ at temperature 0 K. This means, that CO and OC do not arrange quite random in the lattice points. A certain disorder is observed in structure of the crystal.

Problem № 3.33

When $T \to 0$, entropy change of chemical reactions: $\Delta S_{chem.reaction} \to 0$. The reaction: $2C_{graphite} + O_2 = 2CO$ is exception from this rule.
Explain the reason of this phenomenon.

Solution:

Entropy of all substances near the absolute zero approaches zero value: $S \to 0$. According to expression (3.37), ΔS of reaction also approaches to zero: $\Delta S_{chem.react.} \to 0$. But CO is characterized by so–called "residual" entropy, which is conditioned by approximately equal probability of arrangment of CO and OC in the crystalline lattice points. Therefore $S_{CO} = =4.6 \ J/K \times mol$ at 0 K (see problem № 3.32). As a result entropy change of reaction :

$$2C + O_2 \to 2CO$$

is equal to

$$\Delta S = 2S_{CO} - 2S_C - S_{O_2} = 2 \times 4.6 - 0 - 0 = 9.2 \ J/K \times mol$$

Problem № 3.34

Calculate the difference in molar entropies of water and ice under 1 atm pressure and $-10°C$ temperature. Take into consideration, that molar enthalpy of crystallization of water at 0°C is -6.009 kJ / mol, $\overline{C}_{P(H_2O)}$ $=75.50$ J / K×mol and $\overline{C}_{P(ice)} = 36.87$ J / K×mol.

Solution:

State of supercooled water is metastable and therefore water glaciates easily:

$$water_{(-10^0C)} \rightarrow ice_{(-10^0C)} \tag{I}$$

Entropy change in this process is:

$$\Delta\overline{S}_1 = \overline{S}_{ice(-10^0C)} - \overline{S}_{wat.(-10^0C)} \tag{a}$$

(I) process proceeds irreversibly. In order to establish the value of its ΔS, we refer the following way: Let us consider a sequence of reversible processes, in which the initial and final states of system are identical with (I) process:

$$water(-10^0C)\xrightarrow{(1)} water(0^0C)\xrightarrow{(2)} ice(0^0C)\xrightarrow{(3)} ice(-10^0C) \text{ (II)}$$

Following entropy additivity principle,

$$\Delta\overline{S}_{II} = \Delta\overline{S}_1 + \Delta\overline{S}_2 + \Delta\overline{S}_3 \tag{b}$$

The initial and final states of system are the same in the (I) and (II) processes. Since entropy represents a state function, it may be written:

$$\Delta\overline{S}_I = \Delta\overline{S}_{II} \tag{c}$$

Let us calculate entropy change at the constituent stages of (II) process. Water is heated from -10^0C to 0^0C in the (1) process. According expression (3.28):

$$\Delta \overline{S}_{(1)} = \int_{263}^{273} \overline{C}_{p(H_2O)} \frac{dT}{T} = \overline{C}_{p(H_2O)} \ln \frac{273}{263} = 75.5 \times 3.73 \times 10^{-2} =$$

$$= 2.82 \ J / K \times mol$$

(2) process represents the phase transfer (crystallization) at 0^0C. According to (3.25),

$$\Delta \overline{S}_{(2)} = \frac{\Delta \overline{H}_{cryst.}}{T} = -\frac{6009}{273} = -22.01 \ J / K \times mol$$

The cooling of ice from 0^0C to -10^0C takes place in process (3):

$$\Delta \overline{S}_{(3)} = \int_{273}^{263} \overline{C}_{p(ice)} \frac{dT}{T} = \overline{C}_{p(ice)} \ln \frac{263}{273} = 36.87 \times (-3.73 \times 10^{-2}) =$$

$$= -1.38 \ J / K \times mol$$

According to expression (b):

$$\Delta \overline{S}_{II} = 2.82 - 22.01 - 1.38 = -20.57 \ J / K \times mol$$

At the same time,

$$\Delta \overline{S}_{II} = \Delta \overline{S}_{I} \tag{c}$$

Thus

$$\Delta \overline{S}_{I} = \overline{S}_{ice(-10^0 C)} - \overline{S}_{wat.(-10^0 C)} = -20.57 \ J / K \times mol$$

Hence

$$\overline{S}_{wat.(-10^0 C)} - \overline{S}_{ice(-10^0 C)} = 20.57 \ J / K \times mol$$

Thus, molar entropy of water exceeds molar entropy of ice. It must even be so, because crystalline ice represents more ordered structure than liquid water.

Problem № 3.35

Use data and results of problem № 3.34 and calculate entropy changes of surroundings and "universe" at crysrallization of supercooled water. What phase transfer (solidification or melting) does proceed spontaneously in the given conditions (-10^0C, 1 atm)?

Solution:

In the previous problem it was shown, that entropy change of system in the crystallization process of supercooled water is: $\Delta \overline{S}_{syst.} = -20.57 \, J \, / \, K \times mol.$ If judge by the results obtained, the process (I) must not proceed spontaneously:

$$water_{(-10^0 C)} \rightarrow ice_{(-10^0 C)} \tag{I}$$

But it is known from experience, that supercooled water transfers into the crystalline state easily. The reason of this "contradiction" consists in following: entropy increase of system represents a criterion of spontaneity in isolated systems only. Entropy increase of "universe" represents such criterion in nonisolated systems (system considered by us belongs to them):

$$\Delta S_{"universe"} = \Delta S_{syst.} + \Delta S_{surr.} > 0$$

Let us calculate the values of $\Delta S_{surr.}$ and $\Delta S_{"univ."}$.

According to condition of problem № 3.34, molar enthalpy of water crystallization at 0^0C $\Delta \overline{H}_{cryst.} = -6009 \, J \, / \, mol.$ Kirchhoff's law (2.21) is used in order to establish the value of $\Delta \overline{H}_{cryst.}$ at -10^0C:

$$\Delta \overline{H}_{263} = \Delta \overline{H}_{273} + \int_{273}^{263} \Delta \overline{C}_P \, dT \tag{a}$$

Let us consider, that $\Delta\overline{C}_p = const$ in the given temperature range. then it is obtained from (a):

$$\Delta\overline{H}_{263} = \Delta\overline{H}_{273} + \Delta\overline{C}_p\Delta T \qquad\qquad (b)$$

$$\Delta\overline{C}_p = \overline{C}_{p(ice)} - \overline{C}_{p(wat.)} = 36.87 - 75.50 = -38.63\, J/K \times mol$$

$$\Delta T = 263 - 273 = -10\, K$$

Let us introduce these values into expression (b):

$$\Delta\overline{H}_{263} = -6009 + (-38.63)(-10) = -6009 + 386.3 = -5622.7\, J/mol$$

According (3.44),

$$\Delta S_{surr.} = \frac{q_{surr.}}{T} = -\frac{q_{syst.}}{T}$$

In our problem $q_{syst.} = \Delta\overline{H}_{cryst.(H_2O)}$. At 263 K and 1 atm pressure $\Delta\overline{H}_{cryst.(H_2O)} = -5622.7\, J/mol$

Hence

$$\Delta S_{surr.} = -\frac{q_{syst.}}{T} = -\frac{\Delta\overline{H}_{cryst.(H_2O)}}{T} = -\frac{-5622.7}{263} = 21.38\, J/K \times mol,$$

$$\Delta S_{"univ."} = \Delta S_{syst.} + \Delta S_{surr.} = -20.57 + 21.38 = 0.82\, J/K$$

Thus, crystallization of supercooled water proceeds spontaneously. In spite of entropy of system decreases due to formation of more ordered structure (ice), the process is still spontaneous, because entropy increase of surroundings exceeds entropy reducing of system. As a result entropy of "universe" increases and this represents a motive force of the process.

Problem № 3.36

What is the difference between molar entropies of water and its vapor at 90^0C temperature and 1 atm pressure? Take into account, that molar enthalpy of water evaporization at 100^0C is equal to 40,7 kJ/mol , $\overline{C}_{p(H_2O)} = 75.5$ J / K×mol and $\overline{C}_{p(vapor)} = 33.56$ J / K×mol.

Solution:

Consider water vaporization at 90^0C:

$$water_{(90°C)} \rightarrow vapor_{(90°C)} \qquad (I)$$

$$\Delta\overline{S}_I = \overline{S}_{vap.(90^0C)} - \overline{S}_{wat.(90^0C)} \qquad (a)$$

Let carry out (I) process stepwise and reversibly:

$$water(90°C) \xrightarrow{(1)} water(100°C) \xrightarrow{(2)} vapor(100°C) \xrightarrow{(3)} vapor(90°C)$$

$$(II)$$

Due to entropy additivity:

$$\Delta\overline{S}_{II} = \Delta\overline{S}_1 + \Delta\overline{S}_2 + \Delta\overline{S}_3 \qquad (b)$$

According expressions (3.28) and (3.29), entropy change of 1 mole of water at the heating from 90^0C to 100^0C is equal:

$$\Delta\overline{S}_1 = \int_{363}^{373} \overline{C}_p \frac{dT}{T} = \overline{C}_{p(wat.)} \ln\frac{373}{363} = 75.5 \times 2.72 \times 10^{-2} = 2.05 \ J/K \times mol$$

By (3.25), in the evaporization process of water at 100^0C:

$$\Delta\overline{S}_2 = \frac{\Delta\overline{H}_{evap}}{T} = \frac{40700}{373} = 109.12 \ J/K \times mol$$

Entropy change at vapor cooling from 100^0C to 90^0C is equal:

$$\Delta \overline{S}_3 = \int\limits_{373}^{363} \overline{C}_{p(vap.)} \, \frac{dT}{T} = \overline{C}_{p(vap.)} \, \ln \frac{363}{373} = 33.56 \times (-2.72 \times 10^{-2}) = -0.91 \, J \, / \, K \cdot mol$$

According (b),

$$\Delta \overline{S}_{II} = \Delta \overline{S}_1 + \Delta \overline{S}_2 + \Delta \overline{S}_3 = 2.05 + 109.12 - 0.91 = 110.26 \, J \, / \, K \times mol$$

The initial and final states of system in processes (I) and (II) are the same. Therefore

$$\Delta \overline{S}_I = \Delta \overline{S}_{II}$$

Hence

$$\Delta \overline{S}_I = \overline{S}_{vapor.(90°C)} - \overline{S}_{waert(90°C)} = 110.26 \, J \, / \, K \times mol$$

and

$$\overline{S}_{water(90°C)} - \overline{S}_{vapor(90°C)} = -110.26 \, J \, / \, K \times mol$$

Thus, molar entropy of water is less than molar entropy of vapor. It is remarkable, that entropy values of liquid and solid states are much greater neighbouring $(\overline{S}_{water} - \overline{S}_{solid} \approx 20 \, J \, / \, K \times mol)$, than entropies of liquid and vapor states $(\overline{S}_{vapor} - \overline{S}_{water} \approx 110 \, J \, / \, K \times mol)$. This is conditioned by practically structureless character of vapor, whereas liquid (and all the more solid) substance is significantly structurized.

Problem № 3.37
Calculate entropy change of surroundings and "universe" at the evaporation of water at temperature $90°C$ and pressure 1 atm. Use the data and results of problem № 3.36.

Solution:

Let's consider the process of evaporation of water at $90°C$ and 1 atm pressure:

$$water\ (90°C) \to vapor(90°C) \quad (I)$$

$$\Delta \overline{S}_{syst.} = 110.26\ J/K \times mol \qquad \text{(See problem № 3.36)}$$

If we discuss according to this result, the evaporation of water at $90°C$ and 1 atm should proceed spontaneously. However, in nonisolated system the condition of spontaneity of the process is not $\Delta S_{syst.} > 0$, but $\Delta S_{"univ."} > 0$. Therefore let's find $\Delta S_{surr.}$ and $\Delta S_{"univ."}$.

By expression (3.44),

$$\Delta S_{surr.} = \frac{q_{surr.}}{T} = -\frac{q_{syst.}}{T} = -\frac{\Delta \overline{H}_{evap.}}{T}$$

For determination of $\Delta \overline{H}_{evap.}$ at $90°C\,(363K)$ and 1atm pressure Kirchhoff's law is used:

$$\Delta \overline{H}_{363} = \Delta \overline{H}_{373} + \int_{373}^{363} \Delta \overline{C}_p dT \quad (a)$$

Since the temperature range is small, then may be thought that $\Delta \overline{C}_p = const.$ From (a) it is obtained:

$$\Delta \overline{H}_{363} = \Delta \overline{H}_{373} + \Delta \overline{C}_p \cdot \Delta T \quad (b)$$

According to the condition of problem № 3.36, $\Delta \overline{H}_{373} = 40700\ J/mol$

$$\Delta \overline{C}_p = \overline{C}_{p(vap)} - \overline{C}_{p(wat)} = 33.56 - 75.50 = -41.94\ J/K \times mol$$

Let enter these values into expression (b):

$$\Delta \overline{H}_{363} = 40700 + (-41.94) \times (-10) = 40700 + 419.4 = 41119.4 \ J/mol.$$

Then

$$\Delta \overline{S}_{surr.} = -\frac{\Delta \overline{H}_{evap.}}{T} = -\frac{41119.4}{363} = -113.28 \ J/K \times mol,$$

$$\Delta S_{"univ."} = \Delta S_{syst.} + \Delta S_{surr.} = 110.26 - 113.28 = -3.02 \ J/K \times mol.$$

Thus, $\Delta S_{"universe"} < 0$. This means, that the evaporation of water at $90°C$ and 1 atm does not proceed spontaneously. The reversed process – the condensation of vapor into liquid is spontaneous in these conditions.

Remark: Entropy change of system for the phase transformations: $water \xrightarrow{-10°C} ice$ and $water \xrightarrow{90°C} vapor$ considered in problems №№ 3.34 ÷ 3.37 does not represent such value, by using of which the direction of process may be determined. Therefore at the discussion of all the processes it should be determined not only $\Delta S_{syst.}$, but also $\Delta S_{surr.}$ and $\Delta S_{"univ."}$.

Problem № 3.38

Heat exchange and work represent a various mode of energy transfer. Neverthereless heat transfer may be considered as work. Substantiate this assumption.

Solution:

The method of so-called "generalized forces" is adopted in thermodynamics (see chapter 1), following which various modes of work are described by one common formula:

$$\delta A_k = F_k dx_k \tag{3.80}$$

where F_k is generalized force, x_k –generalized coordinate, δA_k represents generalized work.

In case of gas expansion from (3.80) will be obtained:

$$\delta A_{\text{exp}} = PdV \qquad (3.81)$$

where P is generalized force, V– generalized coordinate and δA_k represents work of expansion. The essential and sufficient requirement for fulfilment of work is the change of the generalized coordinate – volume ($dV{\neq}0$).

In process of heat transfer temperature (T) plays a part of generalized force and entropy (S) – of generalized coordinate:

$$\delta q = TdS \qquad (3.82)$$

In case of such approach heat (δq) represents a generalized work, which must be fulfilled (i.e. amount of heat, which must be exchanged) for changing of entropy by the value dS.

Problem № 3.39
Does it exist the analogy between elementary acts of expansion and heating?

Solution:
Work and heat exchange according to the method of generalized forces are described by the common formula:

$$\delta A_k = F_k dx_k \qquad (3.80)$$

where F_k is generalized force, x_k – generalized coordinate, δA_k represents generalized work.

At equilibrium the generalized forces are equal on the both sides of control surface:

$$F_{ex} = F_{in}$$

where F_{ex} is generalized force, which affects on the system externally and F_{in} represents generalized force, by which the system effects on surroundings.

The condition for carrying out of elementary act is:

$$F_{ex} = F_{in} + \delta F \qquad (3.83)$$

where δF is infinitesimal amount of generalized force, which can obtain both positive and negative values.

Let us consider elementary act of expansion. Before beginning of process the system and surroundings are in equilibrium state and the generalized forces on the both sides of control surface are equal:

$$P_{in} = P_{ex}$$

If we reduce P_{ex} by infinitesimal value with external effect, then $P_{in} > P_{ex}$ or it is fulfilled the condition:

$$P_{in} = P_{ex} + \delta P \qquad (\delta P > 0) \qquad (3.84)$$

As a result an elementary act of expansion will be carryied out, at proceeding of which infinitesimal work of expansion is fulfilled:

$$\delta A_{exp} = PdV$$

Due to increasing of volume internal pressure of system (P_{in}) decreases and becomes equal to external pressure (P_{ex}), which is invariable at the proceeding of elementary act:

$$P_{in} = P_{ex}$$

Then elementary act of expansion completes and state of equilibrium is balanced.

Let us consider now an elementary act of heat transfer, when the system receives heat from surroundings. The condition of equilibrium between system and surroundings consists in the equality of generalized forces on the both sides of control surface:

$$T_{in} = T_{ex}$$

If temperature of surroundings exceeds temperature of system by infinitesimal value:

$$T_{ex} = T_{in} + \delta T \qquad (\delta T > 0) \tag{3.85}$$

then infinitesimal difference in generalized forces of system and surroundings (δT) causes the transfer of infinitesimal amount of heat (δq) from surroundings into system. As a result an elementary act of heat exchange is performed between system and surroundings:

$$\delta q = TdS$$

This process is accompanied by infinitesimal change of generalized coordinate – entropy.

An increase of temperature of system (T_{in}) takes place at heat transfer. An equilibrium is balanced, when temperatures of system and surroundings become equal:

$$T_{in} = T_{ex}$$

Thus, it is clear, that a close analogy exists between heat transfer and gas expansion. This is mainly revealed in mechanism of action of generalized forces (T and P).

The similarity between heat exchange and work is stipulated by fact, that both represent the modes of energy transfer. However these modes are dissimilar: at heat transfer energy occures by chaotic and disordered manner, but at fulfilment of work energy transfer is realized by ordered and organized form.

Problem № 3.40

Does entropy change as a result of fulfilment of work? What is the difference between heat and work?

Solution:

As is seen in problem № 3.38, heat transfer may be considered as work, by fulfilment of which entropy changes. The natural question is arisen: does entropy change as a result of performance of "real" work e.g. work of expansion? (We do not concern the irreversible processes, where a part of work is consumed with entropy "formation"). Let us consider for example a reversible adiabatic expansion. In spite of the fact, that work is fulfilled in this process $A = n\overline{c}_v (T_2 - T_1)$, entropy change does not take place. Moreover, $\Delta S = 0$ not only in system, but also in surroundings and "universe" (see problem № 3.23). In some other process both entropy change and fulfilment of work may be occurred, but they do not conditioned each other, e.g. at reversible isothermal expansion both entropy change and fulfilment of work proceed on the basis of heat obtained from reservoir (see problem № 3.24). As concerning expansion in a vacuum, work is not fulfilled here, but entropy increases considerably (problem № 3.26).

The motive force of spontaneous process is an affinity for dissipation of energy. At expansion of gas this is an affinity of energy–possessing particles for occupation of more volume. Work accompanies this phenomenon. It may be fulfilled or not, depending upon the fact exists or not in the system a device by using of which the fulfilment of work is possible, e.g. device with mobile piston of the certain mass. Such device is functionning at reversible isothermal and adiabatic expansions and work is fulfilled. But such device is absent in the system at expansion in a vacuum and therefore work is equal to zero.

Work is related with changes of heights of energy levels of system. When system fulfils a work, the lowering of its energy levels occurs, but when work is fulfilled on system, the rising of energy levels takes place. In this case the number and population of levels (arrangement of particles on these levels) remains invariable.

Entropy change exactly means the change of number of energy levels. The assimilation of even more levels and more uniform distribution of particles, presented in the system, occur with entropy increase. Entropy decrease means reducing of number of assimilated levels and dissimilar redistribution of particles. In this case the heights of energy levels remain invariable. As regards heat exchange, its essence is the same as entropy: the assimilation of even more levels and more uniform redistribution of particles take place at receiving of heat by system, but vice versa at releasing of heat.

Thus, the internal relation between entropy and work does not exist. Therefore, the fulfilment of work in reversible process will not cause the entropy change. In contrast to this, the closest relation exists between entropy and heat (remember if only Clausius's inequality).

Problem № 3.41

Equilibrium heat transfer proceeds by feeding of infinitesimal amount of energy; infinite great time is needed for transferring of finite amount of heat. It is ibvious, that carrying out really equilibrium process is impossible. But we may approach it by performing a heat transfer with minute portions of energy.

Imagine, that 1 mole of helium is heated step–wise from 320 K to 350 K. It receives 0.5 J energy at each step. $\bar{c}_v = 12.47$ J /K×mol for helium.

1) How does change temperature of steps?

2) How does change entropy on first three and final steps?

3) How does express and what equals entropy change at heating from 320 to 350 K?

4) How much heat receives system?

Solution:

1) Due to heat transfer temperature of system at the end of each step increases slightly. Increment of temperature may be calculated by using expression (1.15):

$$q_v = n\bar{c}_v \Delta T$$

Hence,

$$\Delta T = \frac{q_v}{n\bar{c}_v} \qquad (3.86)$$

If the same amount of heat is delivered on all steps, then

$$\Delta T = \frac{0.5}{1 \times 12.47} = 0.04\,K$$

It follows that at the end of each step

$$T_{fin.} = T_{init.} + 0.04$$

2) Since increment of temprature is very minor, it may be considered that on all steps $T = T_{init.}$, which gives the right to use expression (3.21):

$$\Delta S = \frac{q}{T_{init.}}$$

a) First step. $T_{init.} = 320$ K

$$\Delta S_I = \frac{q}{T_{init.}} = \frac{0.5}{320} = 1.5625 \times 10^{-3} \ J \ / \ K \times mol$$

$$T_{fin.} = T_{init.} + 0.04 = 320 + 0.04 = 320.04 \ K$$

b) Second step. $T_{init} = 320.04$ K

$$\Delta S_{II} = \frac{0.5}{320.04} = 1.5623 \times 10^{-3} \ J \ / \ K \times mol$$

$$T_{fin.} = 320.04 + 0.04 = 320.08 \ K$$

c) Third step $T_{init.} = 320.08$ K

$$\Delta S_{III} = \frac{0.5}{320.08} = 1.5621 \times 10^{-3} \ J \ / \ K \times mol$$

$$T_{fin.} = 320.08 + 0.04 = 320.12 \ K$$

d).Final step

$$T_{init.} = 350 - 0.04 = 349.96 \ K$$

$$\Delta S_{fin.st.} = \frac{0.5}{349.96} = 1.4287 \times 10^{-3} \ J \ / \ K \times mol$$

Thus, the higher is initial temperature of step, the smaller is entropy change.

3) Entropy change of system at heating from 320 to 350 K is equal to the sum of ΔS of all steps:

$$\Delta S = \sum \Delta S_k,$$

where k is number of steps.

The sum may be substituted by integral:

$$\Delta S = \int_{T_1}^{T_2} dS = \int_{T_1}^{T_2} \frac{\delta q_v}{T} = \int_{T_1}^{T_2} n\overline{c}_v \frac{dT}{T} = n\overline{c}_v \ln \frac{T_2}{T_1} = 1 \times 12.47 \ln \frac{350}{320} =$$

$$= 1.12 \, J \, / \, K \times mol$$

4) According to expression (1.15), heat received by system is equal:

$$q_v = n\overline{c}_v (T_{fin.} - T_{init.}) = 1 \times 12.47(350 - 320) = 374.1 \, J \, / \, mol$$

Problem № 3.42

Temperature of 1 mole helium is 320 K. 0.5 J heat is transferred from system to surroundings in conditions of constant volume.

1) What became temperature of system?

2) What are equal ΔS and $\Delta \sigma$ of process (use Clausius's and Sacure—Tetrode's equations).

Solution:

1) According to expression (3.86):

$$\Delta T = \frac{q_v}{n\overline{c}_v} = \frac{-0.5}{1 \times 12.47} = -0.04 \, K$$

$$T_{fin.} = T_{init.} - 0.04 = 320 - 0.04 = 319.96 \, K$$

2) Following Clausius's equation (3.21):

$$\Delta S = \frac{q}{T_{init.}} = \frac{-0.5}{320} = -1.56 \times 10^{-3} \; J/K$$

From expression (3.9):

$$\Delta\sigma = \frac{\Delta S}{k} = -\frac{1.56 \times 10^{-3}}{1.38 \times 10^{-23}} = -1.13 \times 10^{20}$$

Let us use Sacure—Tetrode's equation (3.14):

$$\Delta\sigma = N\left(\frac{3}{2}\ln\frac{T_2}{T_1} + \ln\frac{V_2}{V_1}\right) = N\frac{3}{2}\ln\frac{T_2}{T_1} = 6 \times 10^{23} \times 1.5\ln\frac{319.96}{320} =$$

$$= -1.13 \times 10^{20}$$

$$\Delta S = k\Delta\sigma = 1.38 \times 10^{-23}(-1.13 \times 10^{20}) = -1.56 \times 10^{-3} \; J/K$$

Problem № 3.43

Temperature of 1 mole copper changes from 320 to 350 K in isobaric conditions. Molar heat capacity of copper is 25.15 J / K×mol. Consider, that c_p is invariable in the given temperature range and determine:

  1) What amount of heat receives copper?
  2) What are equal ΔS and $\Delta\sigma$ of process?
Compare results obtained in problems №№3.41 and 3.43 with one another.

Solution:

  1) The amount of heat received by copper according equation (1.16) is:

$$q_p = n\bar{c}_p(T_2 - T_1) = 1 \times 25.15(350 - 320) = 754.5 \; J$$

2) According (3.29) entropy change of copper at heating is:

$$\Delta \overline{S} = \int_{T_1}^{T_2} n\overline{c}_p \frac{dT}{T} = n\overline{c}_p \ln \frac{T_2}{T_1} = 1 \times 25.15 \ln \frac{T_2}{T_1} = 25.15 \ln \frac{350}{320} = 2.25 \, J / K \times mol$$

Following (3.9):

$$\Delta \overline{\sigma} = \frac{\Delta \overline{S}}{k} = \frac{2.25}{1.38 \times 10^{-23}} = 1.63 \times 10^{23}$$

Thus, $\Delta \overline{S}_{Cu} = 2.25 \, J / K \times mol$ at change of temperature from 320 to 350 K. It was seen in problem № 3.41, that $\Delta \overline{S}_{He} = 1.12 \, J / K \times mol$ at heating in the same temperature range. Such significant distinction in $\Delta \overline{S}$ of helium and copper is stipulated by different conditions of carrying out of process. Copper heats at constant pressure, but helium at constant volume. In this case $\frac{\overline{c}_{p(Cu)}}{\overline{c}_{v(He)}} = 2$. This means, that isobaric heating of 1 mole Cu needs two–fold more heat, than it is required for heating of 1 mole He in isochoric conditions. Therefore $\Delta \overline{S}_{Cu}$ exceeds $\Delta \overline{S}_{He}$ two–fold.

Problem № 3.44
1 L of ideal gas presents at standard conditions and 298 K. Its pressure increases until 1.7 atm at isochoric heating to 500 K.
    1) By one of Sacure—Tetrode's expressions determine entropy change at heating.
    2) Compare formula obtained by us with expression existing in classical thermodynamics.

Solution:
    1) Let us use expression (3.65):

$$S = kN\left(\frac{5}{2}\ln T - \ln P + const'\right)$$

Hence, in conditions $N = const$

$$\Delta S = S_2 - S_1 = kN\left(\frac{5}{2}\ln\frac{T_2}{T_1} - \ln\frac{P_2}{P_1}\right) \tag{3.87}$$

Taking into account equality $kN = Rn$, from (3.87) is obtained:

$$\Delta S = Rn\left(\frac{5}{2}\ln\frac{T_2}{T_1} - \ln\frac{P_2}{P_1}\right) \tag{3.88}$$

Following ideal gas equation:

$$n = \frac{PV}{RT} = \frac{1 \times 1}{0.082 \times 298} = 0.04\, mol$$

$$\Delta S = 8.31 \times 0.04\left(\frac{5}{2}\ln\frac{500}{298} - \ln\frac{1.7}{1}\right) = 8.31 \times 0.04 \times (1.29 - 0.53) = 0.26\, J/K$$

2) Let us compare expression (3.88) obtained from Sacure—Tetrode's equation with expression (3.32) adopted in classical thermodynamics:

$$\Delta S = Rn\left(\frac{5}{2}\ln\frac{T_2}{T_1} - \ln\frac{P_2}{P_1}\right) \tag{3.88}$$

$$\Delta S = n\left(\overline{c}_p\ln\frac{T_2}{T_1} - R\ln\frac{P_2}{P_1}\right) \tag{3.32}$$

Let us equalize the right–hand sides of expressions (3.88) and (3.32) with each other:

$$Rn\left(\frac{5}{2}\ln\frac{T_2}{T_1} - \ln\frac{P_2}{P_1}\right) = n\left(\overline{c}_p\ln\frac{T_2}{T_1} - R\ln\frac{P_2}{P_1}\right)$$

$$\frac{5}{2}R\ln\frac{T_2}{T_1} - R\ln\frac{P_2}{P_1} = \bar{c}_p \ln\frac{T_2}{T_1} - R\ln\frac{P_2}{P_1}$$

$$\frac{5}{2}R\ln\frac{T_2}{T_1} = \bar{c}_p \ln\frac{T_2}{T_1}$$

$$\frac{5}{2}R = \bar{c}_p \qquad\qquad (3.89)$$

As is known, heat capacity of ideal gas in conditions of constant pressure is described by expression (3.89). Thus, equivalent equations exist in classical and statistical thermodynamics for determination of ΔS of heating at $N=const$ conditions.

Problem № 3.45
Solve problem № 3.44 simpler by one of Sacure—Tetrode's equation.

Solution:
Process described in problem № 3.44 proceeds in $V=const$ conditions. Its ΔS we have determined following expression (3.88):

$$\Delta S = Rn\left(\frac{5}{2}\ln\frac{T_2}{T_1} - \ln\frac{P_2}{P_1}\right)$$

It is preferable in case of $V=const$ to use that equation, which contains parameter with constant value (V). Then one term will be excluded from equation and formula becomes simpler. Let us use expression (3.16):

$$\Delta S = Rn\left(\frac{3}{2}\ln\frac{T_2}{T_1} + \ln\frac{V_2}{V_1}\right) = Rn\frac{3}{2}\ln\frac{T_2}{T_1} = 8.31 \times 0.04 \times \frac{3}{2}\ln\frac{500}{298} =$$

$$= 0.26\, J/K$$

Problem № 3.46

The volume of 0.1 mole ideal gas is reduced from 3.28 L to 2.62 L at cooling under P=const conditions. $\bar{c}_p$ of gas is equal to 20.78 J /K×mol.

 1) Determine entropy change at cooling. Assume, that c_v = const.
 2) What are initial and final temperatures of system, if P = 1 atm?

Solution:

 1) Let us use expression:

$$\Delta S = n\bar{c}_v \ln \frac{T_2}{T_1} + nR \ln \frac{V_2}{V_1} \qquad (3.31)$$

From ideal gas equation under *P,n = const* conditions:

$$\frac{T_2}{T_1} = \frac{V_2}{V_1} \qquad (a)$$

Taking into account this, from (3.31) is obtained:

$$\Delta S = n\bar{c}_v \ln \frac{T_2}{T_1} + nR \ln \frac{V_2}{V_1} = n\bar{c}_v \ln \frac{V_2}{V_1} + nR \ln \frac{V_2}{V_1} = n \ln \frac{V_2}{V_1}(\bar{c}_v + R) =$$

$$= n\bar{c}_p \ln \frac{V_2}{V_1} = 0.1 \times 20.78 \ln \frac{2.62}{3.28} = -0.47 \ J/K$$

 2) Let us determine initial and final temperatures of systems:

$$T_1 = \frac{PV_1}{nR} = \frac{1 \times 3.28}{0.1 \times 0.082} = 400 \ K$$

$$T_2 = \frac{PV_2}{nR} = \frac{1 \times 2.62}{0.1 \times 0.082} = 319.5 \ K$$

Problem № 3.47

Two closed thermodynamic systems containing ideal gas are given. Temperature of first system is T_h, but of second system T_c $(T_h > T_c)$. These systems are placed in entire adiabatic jacket, where their thermal contact is performed.

Assume, that the minute (but not infinitesimal) amount of energy is exchanged between systems and derive the expression, which indicates entropy change at heat exchange.

Solution:

Entropy is an additive value and entropy change of combined system is equal to the sum of entropy changes of its consisting systems:

$$\Delta S_{comb.} = \Delta S_h + \Delta S_c$$

where ΔS_h is entropy change of more hot system and ΔS_c represents entropy change of more cold system.

According to conditions of problem, the minute amount of energy is exchanged between systems. Therefore it may be assumed, that heat transfer proceeds at initial temperature of each system, which permits to use the expression (3.21):

$$\Delta S_h = \frac{q_h}{T_h}; \qquad \Delta S_c = \frac{q_c}{T_c}$$

Hence

$$\Delta S_{comb.} = \frac{q_h}{T_h} + \frac{q_c}{T_c} \qquad (a)$$

The combined system is isolated from surroundings. Therefore energy, evolved by one system, must be involved by the other one:

$$- q_h = q_c \qquad (b)$$

$$(q_h < 0, \ q_c > 0)$$

Taking into account (b), it is obtained from (a):

$$\Delta S_{comb.} = \left(\frac{1}{T_c} - \frac{1}{T_h} \right) q \qquad (3.90)$$

where $q = q_c = -q_h > 0$.

Problem № 3.48

The temperatures of two bodies placed in the common adiabatic jacket are distinguished by finite quantity. Does it proceed reversibly the heat transfer from hot body to the cold one?

(Use the expression (3.90), obtained in problem N 3.47).

Solution:

Let sign the temperatures of hot and cold bodies by T_h and T_c respectively. According to (3.90), entropy change of the "combined" system due to heat transfer, is described by equation:

$$\Delta S_{comb.syst.} = \Delta S_h + \Delta S_c = q \left(\frac{1}{T_c} - \frac{1}{T_h} \right)$$

Since $q > 0$ and $\dfrac{1}{T_c} - \dfrac{1}{T_h} > 0$, therefore

$$\Delta S_{comb.syst.} > 0. \qquad (a)$$

Surroundings does not participate in the process of heat transfer according to the condition of problem:

$$\Delta S_{surr.} = 0 \qquad (b)$$

Taking into account (a) and (b), it is obtained:

$$\Delta S_{"univ."} = \Delta S_{comb.syst.} + \Delta S_{surr.} > 0$$

The entropy increase of "universe" indicates, that the process is irreversible (non-equilibrium) and spontaneous. The analogous result is obtained if exchange of infinitesimal amount of energy (δq) between systems is considered.

The analysis of expression (3.90) demonstrates, that non-equilibrium character of the process is conditioned by the stronger change of entropy of the cold body than the hot one during the heat exchange. What may cause it?

The entropy increase in conditions $N=const$ means the occupation of new energy levels and more uniform distribution of particles on the levels. In contrast to this, entropy decrease implies reducing of number of energy levels and less uniform distribution of particles. The more is the temperature, the more levels are already occupied and distribution is also more uniform. Therefore the absorbing or releasing of new portion of energy at high temperature does not induce such sharp changes of both number of levels and distribution of particles, that at lower temperature.

Hence, increasing of entropy of the cold body due to heat exchange is more considerable, than entropy reducing of the hot body. Because of this, entropy of the combined system increases, which indicates, that heat transfer is non-equilibrium process. With that, the more is the difference between temperatures of hot and cold bodies, the more non-equilibrium and spontaneous is the process.

Problem № 3.49

The temperatures of two bodies are distinguished by finite quantity. As we have seen in problem № 3.48, the heat transfer in the combined system, isolated from surroundings, proceeds non-equilibriumly (irreversibly).

Describe the method, by using of which the carrying out of reversible elementary act of heat transfer will be possible.

Solution:

With this purpose let us focus attention on surroundings. Let carry out the process step-wise.

1) Heat transfer from hot body to surroundings.

Let us assume, that the temperature of surroundings is less by infinitesimal than the temperature of hot body. Then infinitesimal amount of heat δq transfers from hot body to surroundings. As result entropy change of hot body is equal:

$$dS_{h(1)} = -\frac{\delta q}{T_h} < 0$$

where $\delta q > 0$ (see problem № 3.47).

The cold body does not participate in this act. Hence

$$dS_{c(1)} = 0$$

and

$$dS_{comb.(syst(1)} = dS_{h(1)} + dS_{c(1)} = -\frac{\delta q}{T_h}.$$

Entropy change of surroundings is:

$$dS_{surr.(1)} = \frac{\delta q}{T_h} > 0$$

Entropy change of "universe" is equal:

$$dS_{"univ."(1)} = dS_{comb.syst.(1)} + dS_{surr.(1)} = -\frac{\delta q}{T_h} + \frac{\delta q}{T_h} = 0$$

2) Expansion of surroundings.

Let expand the surroundings adiabatically and reversibly until its temperature reduces from T_h to T_c. Since the expansion is adiabatic and reversible and the hot and cold bodies do not participate in this process, therefore

$$dS_{surr(2).} = 0, \; dS_{h(2)} = dS_{c(2)} = dS_{comb.syst.(2)} = 0 \quad \text{and} \quad dS_{"univ.'(2)} = 0$$

3) Heat transfer from surroundings to the cold body.

Let give surroundings possibility of thermal contact and transferring of heat δq to the cold body. Entropy change of cold body will be:

$$dS_{c(3)} = \frac{\delta q}{T_c} > 0$$

Since the hot body does not participate in this process, then

$$dS_{h(3)} = 0 \qquad \text{and} \qquad dS_{comb.syst.(3)} = dS_{c(3)} = \frac{\delta q}{T_c}$$

Entropy change of surroundings equals:

$$dS_{surr.(3)} = -\frac{\delta q}{T_c} < 0,$$

And entropy change of "universe" is equal:

$$dS_{"univ."(3)} = dS_{comb.syst.(3)} + dS_{surr.(3)} = \frac{\delta q}{T_c} - \frac{\delta q}{T_c} = 0.$$

Thus, the infinitesimal amount of heat is transferred from hot body to the cold one through surroundings. Let analyze reversibility of this process.

Entropy is the additive quantity. Entropy change of "universe" in the total process is equal to the sum of entropy changes of "universe" on the constituting steps of process:

$$dS_{"univ."} = dS_{"univ."(1)} + dS_{"univ."(2)} + dS_{"univ."(3)} = 0 + 0 + 0 = 0.$$

This means that the heat transfer from the hot body to the cold one, carried out by participating of surroundings, may proceed equilibriumly. Such character of process will be stipulated by infinitesimal difference between temperatures of hot body and surroundings firstly, and between the cold body and surroundings after adiabatic expansion of surroundings.

When the temperatures of two objects are distinguished by infinitesimal, then distribution of particles on the occupied energy levels differs by infinitesimal also. Therefore entropies of objects change by the same value, due to absorbing or releasing of equal amount of heat (δq). (Under the "objects" the hot and cold bodies and surroundings are implied).

Remark: from the considered example is shown, that for carrying out the reversible heat transfer it is necessary to perform work (namely, to perform reversible adiabatic expansion of surroundings). It is interesting, that not only the reversible heat transfer from the cold body to the hot one is impossible (the second law of thermodynamics), but also the reversible heat transfer from the hot body to the cold one is impossible without fulfilling the work. However these two impossibilities are stipulated by the radically different reasons: the heat transfer from cold to the hot body is unnatural process, realization of which is possible only by means of the enforcement. Opposite process – the heat transfer from hot to the cold body is natural process of such extent, that in order to restrain it, the certain enforcement, i.e. performing of work is necessary.

Problem № 3.50

Take into consideration results of problem № 3.49 and derive expressions for entropy change of hot and cold bodies and combined system in the process of reversible heat exchange.

Solution:

As we have seen in problem № 3.49, at the carrying out elementary act of heat transfer by reversible manner:

$$dS_h = -\frac{\delta q}{T_h}, \qquad dS_c = \frac{\delta q}{T_c}$$

$$dS_{comb.} = dS_h + dS_c = \frac{\delta q}{T_c} - \frac{\delta q}{T_h}$$

It is obtained for finite process:

$$\Delta S_h = \int_{T_h}^{T_h'} \frac{\delta q_h}{T} \tag{3.91}$$

$$\Delta S_c = \int_{T_c}^{T_c'} \frac{\delta q_c}{T} \tag{3.92}$$

$$\Delta S_{comb.} = \int_{T_c}^{T_c'} \frac{\delta q_c}{T} + \int_{T_h}^{T_h'} \frac{\delta q_h}{T} \tag{3.93}$$

where T_h' and T_c' are temperatures of more hot and more cold bodies after heat exchange.

In case of thermal equilibrium $T_h' = T_c' = T_{equil.}$ and from (3.93) is obtained:

$$\Delta S_{comb.} = \int_{T_c}^{T_{equil.}} \frac{\delta q_c}{T} + \int_{T_h}^{T_{equil.}} \frac{\delta q_h}{T} \tag{3.94}$$

Problem № 3.51

Two closed thermodynamic system are given. Each system contains 0.05 mole of helium. Temperature of first and second systems are 350 K and 290 K respectively. 15×10^{-3} J energy is transferred from first system to the second one, after which thermal contact is interrupted. $\bar{c}_v$ of helium is equal to 12.47 J / K×mol.

1) What are temperatures of each system at the end of process?
2) What is equal entropy change at heat exchange?

Solution:

1) The process proceeds in conditions of constant volume, therefore energy exchanged between systems is equal to:

$$q_v = n\overline{c}_v (T_{fin.} - T_{init.})$$ (1.15)

It follows that

$$T_{fin.} = \frac{q_v}{n\overline{c}_v} + T_{init.}$$ (a)

Hoter system releases energy: $q_h = -15 \times 10^{-3}$ $J..T_{init.} = T_h = 350$ K. According to (a), the final temperature of hoter system is:

$$T_h' = \frac{-15 \times 10^{-3}}{0.05 \times 12.47} + 350 = -0.024 + 350 = 349.976 \ K$$

Colder system receives energy: $q_c = 15 \times 10^{-3}$ $J.$ $T_{init.} = T_c = 290$ K. Its final temperature is:

$$T_c' = \frac{15 \times 10^{-3}}{0.05 \times 12.47} + 290 = 0.024 + 290 = 290.024 \ K$$

2) As is seen, temperature of both systems due to heat transfer changes negligible ($\Delta T = \pm 0.024$ K). Therefore it may be assumed that process in each system proceeds at initial temperature. This assumption gives us a right to use expression (3.90):

$$\Delta S_{comb.} = \left(\frac{1}{T_c} - \frac{1}{T_h} \right) q_v = \left(\frac{1}{290} - \frac{1}{350} \right) \times 15 \times 10^{-3} =$$

$$5.91 \times 10^{-4} \times 15 \times 10^{-3} = 8.87 \times 10^{-6} \ J/K$$

Remark: It is obtained by using of more precise expression (3.93):

$\Delta S_{comb.} = 8.73 \times 10^{-6} \, J/K$. As is seen, the difference is not great (~1.5%).

Problem № 3.52

Two closed thermodynamic systems are localized in entire adiabatic jacket. Each system contains 0.05 M of helium. Temperatures of hot and cold systems are equal to 350 K and 290 K respectively. The systems are brought into thermal contact until the equalization of temperatures.

1) What is the equilibrium temperature of combined system?

2) What is the energy given off by hot system and received by cold system?

Solution:

1) Energy released by hot system according to expression (1.15) is equal:

$$q_h = n\bar{c}_v (T_{equil.} - T_h)$$

Energy received by cold system is:

$$q_c = n\bar{c}_v (T_{equil.} - T_c)$$

where T_c and T_h are initial temperatures of hot and cold systems respectively.

The combined system is isolated from surroundings with adiabatic jacket. Therefore energy released by hot system is equal to energy received by cold system:

$$-q_h = q_c$$

that is

$$-n\bar{c}_v (T_{equil.} - T_h) = n\bar{c}_v (T_{equil.} - T_c)$$

$$-T_{equil.} + T_h = T_{equil.} - T_c$$

$$T_{equil.} = \frac{T_h + T_c}{2} = \frac{350 + 290}{2} = 320\,K$$

2) Energy released by hot system is:

$$q_h = n\bar{c_v}(T_{equil.} - T_h) = 0.05 \times 12.47(320 - 350) = -18.71\,J$$

Energy received by cold system is equal:

$$q_c = n\bar{c_v}(T_{equil.} - T_c) = 0.05 \times 12.47(320 - 290) = 18.71\,J$$

Problem № 3.53

How will change entropy in the combined system, described in problem № 3.52 after establishment the thermal equilibrium? Assume, that $\bar{c_v}$ of helium is equal to 12.47 J / K×mol in temperature range 290÷350 K.

Solution:

Entropy change in the combined system after reaching the thermal equilibrium is expressed so:

$$\Delta S_{comb.} = \int\limits_{T_h}^{T_{equil.}} \frac{\delta q_h}{T} + \int\limits_{T_c}^{T_{equil.}} \frac{\delta q_c}{T} \tag{3.94}$$

As is known

$$\delta q_v = n\bar{c_v}dT \tag{1.15}$$

Let us introduce expression (1.15) into (3.94). It is obtained:

$$\Delta S_{comb.} = \int\limits_{T_h}^{T_{equil.}} n\bar{c_v}\frac{dT}{T} + \int\limits_{T_c}^{T_{equil.}} n\bar{c_v}\frac{dT}{T} = n\bar{c_v}\left(\ln\frac{T_{equil.}}{T_h} + \ln\frac{T_{equil.}}{T_c} \right) =$$

$$= 0.05 \times 12.47 \left(\ln \frac{320}{350} + \ln \frac{320}{290} \right) = 0.624(-8.96 \times 10^{-2} + 9.84 \times 10^{-2}) =$$

$$= 5.5 \times 10^{-3} \ J/K$$

Problem № 3.54

100 g and 200 g of water with temperatures $80^0 C$ and $40^0 C$ respectively are mixed in the isolated vessel. $\overline{c}_p$ of water is equal to 75.5 J / K×mol.

1) What is the final temperature of system?

2) What is the entropy change in this process?

Solution:

1) The amount of exchanged heat in conditions of $P = const$ is:

$$q_p = n \overline{c}_p (T_{equil.} - T_{init.}) \tag{1.16}$$

Since mixing occurs in the isolated vessel, heat given off by water with temperature $80^0 C$ equals to heat received by water with $40^0 C$:

$$- n_h \overline{c}_p (T_{equil.} - T_h) = n_c \overline{c}_p (T_{equil.} - T_c)$$

where letter symbols "*h*" and "*c*" correspond to hoter and colder water respectively.

$$- n_h (T_{equil.} - T_h) = n_c (T_{equil.} - T_c)$$

$$- n_h T_{equil.} + n_h T_h = n_c T_{equil.} - n_c T_c$$

$$(n_h + n_c) T_{equil.} = n_h T_h + n_c T_c$$

$$T_{equil.} = \frac{n_h T_h + n_c T_c}{n_h + n_c} \tag{3.95}$$

According condition of problem, $T_h = 80^0C = 353K$, $T_c = 40^0C = 313K$,

$$n_h = \frac{100}{18} = 5.55 \; moles, \; n_c = \frac{200}{18} = 11.11 \; moles.$$

Let us untroduce these values into expression (3.95). We obtain:

$$T_{equil.} = \frac{5.55 \times 353 + 11.11 \times 313}{5.55 + 11.11} = \frac{5436.58}{16.66} = 326.3 \; K$$

2) Mixing and heat exchange occur simultaneously in the considered process. Entropy is an additive quantity and change of total entropy equals:

$$\Delta S = \Delta S_{mix.} + \Delta S_{heat\;exch.}$$

$\Delta S_{mix.} = 0$ at mixing of different amounts of the same substance (see problem № 3.30). Hence

$$\Delta S = \Delta S_{heat\;exch.} = \Delta S_h + \Delta S_c$$

where ΔS_h is entropy change of more hot system and ΔS_c represents entropy change of more cold system.

According to (3.91) and (3.92):

$$\Delta S_h = \int_{T_h}^{T_{equil.}} dS_h = \int_{T_h}^{T_{equil.}} \frac{\delta q_h}{T} = \int_{T_h}^{T_{equil.}} n_h \bar{c}_p \frac{dT}{T} = n_h \bar{c}_p \ln \frac{T_{equil.}}{T_h} =$$

$$= 5.55 \times 75.5 \ln \frac{326.3}{353} = -32.92 \; J/K$$

$$\Delta S_c = n_c \bar{c}_p \ln \frac{T_{equil.}}{T_c} = 11.11 \times 75.5 \ln \frac{326,3}{313} = 34.98 \; J/K$$

$$\Delta S = \Delta S_h + \Delta S_c = -32.92 + 34.98 = 2.06 \; J/K$$

Problem № 3.55

Two same rods of copper with weight 500 g are given. Their temperature are 300 and 600 K respectively. They are put in contact till thermal equilibrium.

1) What is entropy change in this process, if assume that molar heat capacity of copper in the given temperature range is 25.15 J /K ×mol?

2) Do heat transfer from hot body to cold and a process of equalization of temperatures proceed spontaneously or not?

Solution:

1) Entropy is an additive quantity and entropy change of combined system is equal:

$$\Delta S_{comb.} = \Delta S_h + \Delta S_c \qquad (a)$$

where ΔS_h and ΔS_c are entropy changes of more hot and more cold bodies respectively.

According expressions (3.91) and (3.92), in the conditions of constant pressure:

$$\Delta S_h = n_h \bar{c}_p \ln \frac{T_{equil.}}{T_h} \qquad (b)$$

$$\Delta S_c = n_c \bar{c}_p \ln \frac{T_{equil.}}{T_c} \qquad (c)$$

From the condition of problem, $n_h = n_c = n$. Then it is obtained from expressions (a), (b) and (c):

$$\Delta S_{comb.} = n\bar{c}_p \left(\ln \frac{T_{equil.}}{T_h} + \ln \frac{T_{equil.}}{T_c} \right) = n\bar{c}_p (2 \ln T_{equil.} - \ln T_h T_c) =$$

$$= n\bar{c}_p (\ln T_{equil.}^2 - \ln T_h T_c) \qquad (d)$$

Following expression (3.95),

$$T_{equil.} = \frac{n_h T_h + n_c T_c}{n_h + n_c}$$

When $n_h = n_c = n$, then

$$T_{equil.} = \frac{n(T_h + T_c)}{2n} = \frac{T_h + T_c}{2} \qquad (e)$$

Let us introduce expression (e) into (d):

$$\Delta S_{comb.} = n\overline{c}_p (\ln T_{equil.}^2 - \ln T_h T_c) = n\overline{c}_p \left[\ln \left(\frac{T_h + T_c}{2} \right)^2 - \ln T_h T_c \right] =$$

$$= n\overline{c}_p \ln \left[\frac{(T_h + T_c)^2}{4 T_h T_c} \right]$$

The number of moles of copper is: $n = \dfrac{500}{63.5} = 7.87 \; moles$ and

$$\Delta S_{comb.} = 7.87 \times 25.15 \ln \left[\frac{(600 + 300)^2}{4 \times 600 \times 300} \right] = 7.87 \times 25.15 \times 0.12 = 23.31 J / K$$

2) The condition must be realized at spontaneous proceeding of process:

$$\Delta S_{"univ."} = \Delta S_{comb.} + \Delta S_{surr.} > 0$$

According to condition of problem, surroundings do not participate in the process of heat exchange. Therefore $\Delta S_{surr.} = 0$. Thus, process proceeds spontaneously, when

$$\Delta S_{"univ."} = \Delta S_{comb.syst.} = n\overline{c}_p \ln \left[\frac{(T_h + T_c)^2}{4 T_h T_c} \right] > 0,$$

or when

$$\frac{(T_h + T_c)^2}{4T_h T_c} > 1.$$

Let us determine correctness of inequality:

$$(T_h + T_c)^2 > 4T_h T_c$$

$$T_h^2 + 2T_h T_c + T_c^2 > 4T_h T_c$$

$$T_h^2 - 2T_h T_c + T_c^2 > 0$$

$$(T_h - T_c)^2 > 0$$

Since the raising of any expression to the second power gives the positive number, the considered inequality is correct. This means, that both heat transfer from hot body to cold one and equalization of temperatures proceed spontaneously. The reversed process – heat transfer from cold body to hot one and formation of temperature gradient in system, presented in thermal equilibrium, do not proceed spontaneously. In order to carry out such process fulfilment of certain work is necessary.

Problem № 3.56

Two same rods of copper with different temperature are given. They relate with one another by wire, which possesses the infinitesimal heat conductivity. Therefore heat transfer:

hot body→wire→cold body

proceeds infinitely slowly.
    Is this process reversible?

Solution:

The process is reversible, if return of system to the initial state is possible via the same way; in this case no change must remain in the surroundings. The considered system does not revert the initial state by the same way, since spontaneous heat transfer from the cold body to hot one is impossible.To do this a fulfilment of work is necessary, which causes a certain changes in the surroundings unequivocally (see also problem № 1.18).

Thus, infinitely slow proceeding of process does not mean yet that the process is reversible. This is essential but insufficient condition for the reversibility.

Creation of peculiar conditions is needed for carrying out the heat transfer by reversible manner (see problem № 3.49).

Problem № 3.57

Use expression for efficiency of Carnot's direct cycle and indicate, that absolute temperature must not have a negative value.

Solution:

Efficiency of Carnot's direct cycle is:

$$\eta = \frac{T_h - T_c}{T_h} = \frac{A}{q_h} \qquad (3.17)$$

where T_h and T_c are temperatures of heater and refrigerator respectively, q_h represents heat received by system (ideal gas) from heater, on the basis of which system fulfils a work; A is algebraic sum of works fulfilled by system and on the system.

Let us assume, that temperature of refrigerator is negative. Then it is obtained from (3.17):

$$\eta = \frac{T_h - (-T_c)}{T_h} = \frac{T_h + T_c}{T_h} > 1,$$

that is

$$\frac{A}{q_h} > 1.$$

A represents a work fulfilled by system on the basis of heat q_h. A must not exceed q_h according the law of energy constancy and conservation. $A > q_h$ is adopted on the basis of assumption, that $T_c < 0$. And because A must not exceed q_h, the absolute temperature can not be negative.

Problem № 3.58
Use efficiency of Carnot's reversed cycle and indicate, that the attainment an absolute zero of temperature is impossible.

Solution:
Efficiency of Carnot's refrigerating cycle is expressed so:

$$\eta' = \frac{T_c}{T_h - T_c} = \frac{q_c}{A} \qquad (3.19)$$

where T_c and T_h are temperatures of refrigerator and heater respectively, q_c is heat taken from refrigerator, A represents a work which must be fulfilled in order to make possible heat transfer from refrigerator to heater.
It is obtained from (3.19):

$$A = q_c \frac{T_h - T_c}{T_c}$$

According to this expression near an absolute zero, when $T_c \rightarrow 0$, then $A \rightarrow \infty$.

Thus, in order to reach absolute zero, the fulfilment of infinitely great work is required, which is impossible. Therefore absolute zero of temperature is unattainable.

Problem № 3.59

Carnot's direct cycle involves following steps:

I. Isothermal expansion from V_1 volume to V_2;

II. Adiabatic expansion from V_2 to V_3;

III. Isothermal compression from V_3 to V_4;

IV. Adiabatic compression from V_4 to V_1.

Take into consideration, that $\dfrac{V_2}{V_1} = \dfrac{V_3}{V_4}$ and prove that entropy is a state function. Plot indicator– diagram in coordinates *S—T*.

Solution:

It follows from entropy additivity, that

$$\oint dS = \int_1^2 dS + \int_2^3 dS + \int_3^4 dS + \int_4^1 dS \qquad (a)$$

where $\oint dS$ is entropy change in circle process; terms on right-hand side of expression represent entropy change at I, II, III and IV steps.

Adiabatic processes are isoentropic: *S=const* at their proceeding. Hence

$$\int_2^3 dS = \int_4^1 dS = 0$$

and expression (a) is transformed so:

$$\oint dS = \int_1^2 dS + \int_3^4 dS \qquad (b)$$

According expression (3.33), at isothermal change of volume:

$$\int_1^2 dS = R \ln \frac{V_2}{V_1} \tag{c}$$

and

$$\int_3^4 dS = R \ln \frac{V_4}{V_3} \tag{d}$$

Following condition of problem, $V_2 / V_1 = V_3 / V_4$. Then

$$R \ln \frac{V_2}{V_1} = -R \ln \frac{V_4}{V_3} \tag{e}$$

If take into account (c), (d) and (e), it is obtained:

$$\int_1^2 dS = - \int_3^4 dS$$

It follows that

$$\oint dS = \int_1^2 dS + \int_3^4 dS = - \int_3^4 dS + \int_3^4 dS = 0$$

Thus, $\oint dS = 0$. This means, that entropy is a state function.

Let us plot indicator – diagram in S—T coordinates (Fig.3.1).

$1 \rightarrow 2$ Isothermal expansion, which proceeds at temperature of heater T_h. Entropy increases from S_1 to S_2.

$2 \rightarrow 3$ Adiabatic expansion in isoentropic ($S_2 = const$) conditions. Temperature decreases from T_h to T_c.

$3 \rightarrow 4$ Isothermal compression at temperature of refrigerator T_c. Entropy decreases from S_2 to S_1.

$4{\rightarrow}1$ Adiabatic compression in conditions $S_1 = const.$ temperature increases from T_c to T_h.

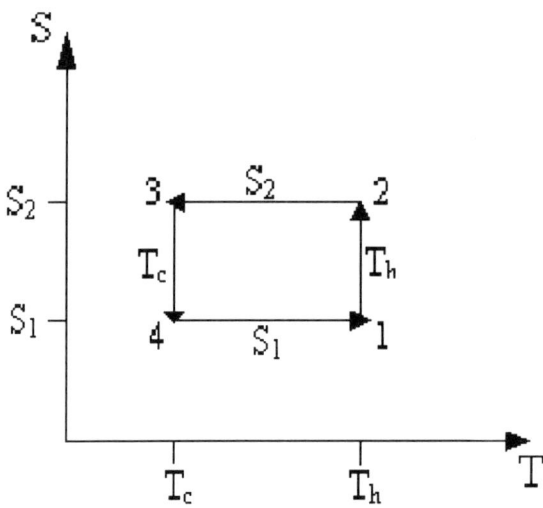

Fig.3.1. Indicator – diagram for Carnot's cycle

Problem № 3.60

Temperature of engine cylinder of helicopter is 2000^0C, but exhaust gases 900^0C. What height may reach the helicopter with weight 2 ton at combustion of 1 gallon benzine, if heat of combustion of benzine is 46.8×10^3 kJ /kg, but density equals to 0.8 g / cm^3? (Take into consideration, that 1 gallon = 3.79 L).

Solution:

A helicopter belongs to thermal mashine working by internal combustion engine. It does not possess an external source of energy (heater). Chemical reaction (combustion of fuel) represents an energy source in helicopter. A fast transformation of chemical energy into thermal energy takes place in engine at a moment of ignition, due to which work is fulfilled and helicopter shifts. As refrigerator the metallic frame or cooling system is used.

Expression for upper limit of efficiency of helicopter is following*:

$$\eta = \frac{T_h - T_c}{T_h} = \frac{A}{q} \tag{3.17}$$

where T_h is a maximum temperature in engine at a moment of ignition, T_c represents temperature of exhaust gases, q is heat obtained at combustion of fuel, but A represents a work fulfilled by helicopter.

Let us determine the amount of heat released as a result of combustion of 1 *gallon* of benzine. 1 *gallon* of benzine occupies 3.79 L volume. According to condition of problem, 3.79 L of benzine weights:

$$0.8 \times 3.79 \times 10^3 = 3.03 \times 10^3 \, g = 3.03 \, kg$$

1 *gallon* (3.03 *kg*) of benzine releases at combustion:

$$q = 3.03 \times 46.8 \times 10^3 = 141.804 \times 10^3 \, kJ \quad \text{heat.}$$

An upper limit of efficiency of helicopter according expression (3.17) is:

$$\eta = \frac{T_h - T_c}{T_h} = \frac{2273 - 1173}{2273} = 0.484 \ (48.4\%)$$

Work fulfilled by helicopter according to the same expression is:

$$A = \eta q = 0.484 \times 141.804 \times 10^6 = 68.633 \times 10^6 \, J = 6.863 \times 10^6 \, kgm$$

$$(1J \approx 0.1 kgm)$$

Let us determine the height which may be reached theoretically by helicopter due to fulfilling of 6.863×10^6 *kgm* work.

$$A = Ph$$

where A is a work, P and h represent weight of helicopter and predicted height respectively.

---

*Efficiency of Carnot's cycle exceeds efficiency of any other cycle working in the same range of temperature. Therefore the establishment of upper limit of efficiency for all thermal machines is carried out by using of Carnot's cycle.

Hence

$$h = \frac{A}{p}$$

According to condition of problem, the helicopter weighs 2 *ton* = 2×10^3 *kg*. Then

$$h = \frac{6.863 \times 10^6}{2 \times 10^3} = 3.43 \times 10^3 \ m = 3.43 \ km$$

Thus, helicopter with weight 2 *ton* and efficiency 48%, by using of 1 *gallon* benzine may reach theoretically the height ~3.4 *km*.

Problem № 3.61

Power of engine of motor-car is 40 horsepower (hp). it consumes 6 L of benzine per hour.

1) What is the the efficiency of engine if combustion heat of benzine is 46.8×10^3 kJ / kg and density equals to 0.8 kg / L?

2) What is the theoretical (thermodynamically permissible) efficiency of engine if temperature of its cylinder is 2000^0C, but temperature of exhaust gases makes up 700^0C?

Solution:

1) $$\eta = \frac{A}{q}$$

where A is work fulfilled by engine and q represents heat released at the combustion of fuel.

Power of engine according to condition of problem is equal:

$$N = 40 \, hp = 40 \times 735 \, J / s = 29400 \, J / s = 29.4 \, kwt$$

Engine will fulfil per hour:

$$A = 29400 \times 3600 = 105 \times 10^6 \ J = 105 \times 10^3 \ kJ \quad \text{work}$$

6 L of benzine weighs: $6 \times 0.8 = 4.8 \ kg$
It is released at combustion of 4.8 *kg* benzine:

$$q = 4.8 \times 46.8 \times 10^3 = 224 \times 10^3 \ kJ \quad \text{heat}$$

$$\eta = \frac{A}{q} = \frac{105 \times 10^3}{224 \times 10^3} = 0.471 \ (47.1\%)$$

2) Theoretical efficiency of motor-car is equal to efficiency of Carnot's direct cycle:

$$\eta_{theor.} = \frac{T_h - T_c}{T_h} = \frac{2000 - 700}{2273} = 0.572 \ (57.2\%)$$

Thus, the difference in thermodinamical and real efficiencies is 10%.

Problem № 3.62
Explain how occurs heat transfer from refrigerator to heater?

Solution:
Let us consider Carnot's reversible refrigerating cycle, where system is a working body and heater and refrigerator represent surroundings. No changes remain in the system and surroundings after completion of cycle because of its reversibility. Accordingly, at the end of circular process:

$$\Delta S_{syst.} = \Delta S_{surr.} = 0$$

As is already noted, *surroundings* = *heater* + *refrigerator*. It follows that

$$\Delta S_{surr.} = \Delta S_{heater} + \Delta S_{refrig.} = 0$$

and

$$\Delta S_{heater} = -\Delta S_{refrig.} \qquad\qquad (a)$$

Let us assume, that heat q_c is taken from refrigerator, temperature of which is T_c. According expression (3.21) entropy of refrigerator decreases:

$$\Delta S_{refrig.} = \frac{q_c}{T_c} < 0 \qquad\qquad (b)$$

Heat is given to heater, temperature of which is T_h. Entropy of heater increases:

$$\Delta S_{heat.} = \frac{q_h}{T_h} > 0 \qquad\qquad (c)$$

It is obtained from expressions (a), (b), (c):

$$\frac{q_h}{T_h} = \frac{q_c}{T_c} \qquad\qquad (d)$$

Since temperature of refrigerator is less than temperature of heater ($T_c < <T_h$), equality (d) will be realized only in case if $q_h > q_c$ i.e. when heater receives more energy, than refrigerator evolves. This is possible only then, when energy (in the form of work) is delivered to heater from the outside:

$$q_h = |q_c| + A \qquad\qquad (3.18)$$

Thus, work of refrigeration plant (heat transfer from cold body to hot one) is possible only with compulsion i.e. by expending of work from the outside.

Problem № 3.63

Temperature of freezed water presented in the chamber of domestic refrigerator is 0^0C, but room temperature equals to 25^0C. What amount of work must be fulfilled in order to take from ice 2 kJ energy? Will the room cool or not due to working of refrigerator?

Solution:

At working of conventional domestic refrigerator air of room is a heater but medium of chamber of domestic refrigerator represents the refrigerator. Heat transfer from refrigerator to heater is not realizable without expensing a work for compression of working body e.g. freon.

Efficiency of Carnot's reversed cycle (so–called cooling coefficient) represents the upper limit of efficiency of refrigeration plant:

$$\eta' = \frac{T_c}{T_h - T_c} = \frac{q_c}{A} \tag{3.19}$$

where T_c is a temperature of chamber of refrigerator (camera), T_h is a temperature of heater (room), q_c represents heat taken from refrigerator and A is a work fulfilled on system from the outside, by means of heat transfer from refrigerator to heater occurs.

According to condition of problem $T_h = T_{room} = 298$ K and $T_c = T_{cam.} = 273$ K

Then

$$\eta' = \frac{T_c}{T_h - T_c} = \frac{273}{25 - 0} = 10.92$$

Following (3.19):

$$\eta' = \frac{q_c}{A}$$

Hence

$$A = \frac{q_c}{\eta'} = \frac{2000}{10.92} = 183.15 \, J$$

Accordingly, if $\sim 183 \, J$ energy is expended in the form of work, then much more energy $\sim 2000 \, J$ will be taken from freezed water. This energy is transferred from back wall of refrigerator to the room. Therefore the room does not cool, but warms. While the increment of temperature is negligible for the room, the back wall of refrigerator obtains a significant amount of energy during working of refrigeration plant.

Problem № 3.64

The temperature of room is $25°C$, but the temperature of surroundings is equal to $35°C$. The conditioner performs 10 kJ work for room cooling.

1) What amount of heat is taken away from the room?
2) What amount of energy is transferred to surroundings?

Solution:

Conditioner represents a cooling plant, where a surroundings is a heater (more hot body) and room appears as a refrigerator (more cold body).

1) The upper limit of the efficiency of the conditioner is equal to the efficiency of Carno's reversed cycle, which works in the same temperature range:

$$\eta' = \frac{T_c}{T_h - T_c} = \frac{q_c}{A}, \qquad (3.19)$$

where T_c is the temperature of refrigerator (room), T_h – temperature of heater (surroundings), q_c represents heat, taken away from the refrigerator and A is work, performed on the system.

According to the condition of problem, $T_h = T_{surr.} = 308K$, $T_c = T_{room} = 298K$.

$$\eta' = \frac{T_c}{T_h - T_c} = \frac{298}{35 - 25} = 29.8$$

On the other hand,

$$\eta' = \frac{q_c}{A} \qquad (3.19)$$

Hence, the heat taken from the refrigerator (room) is equal to:

$$q_c = \eta' \cdot A = 29.8 \cdot 10 = 298 kJ .$$

2) Following expression (3.18), not only the heat taken from the refrigerator (q_c), but also work A, performed on the system, are transferred to the heater. Thus, the total energy transferred to the surroundings is equal to

$$q_{surr.} = q_{room} + A = 298 + 10 = 308 \, kJ .$$

Remark: In the refrigerator plant the energy taken away from the refrigerator (more cold body) exceeds much the work, expended for taking off this energy. For example, in the home refrigerator obtaining of 2000 *J* is possible by expense of 183 *J* work (problem № 3.63); in the conditioner approximately 300 *kJ* energy is produced at the expense of 10 *kJ* work (problem № 3.64).

Working of so–called "heat pump" is also based on refrigerative principle. This plant represents a refrigerator, however, not cooling but the energy obtained at the cooling, which may be used for heating, is topical in this case. The idea about using refrigerator plant in the heating purpose was firstly proposed in 1853. Nowadays widely used conditioner is utilized as a heat pump. This plant behaves as a cooler in summer: takes heat from room (refrigerator) and transfers it to surroundings (heater). But it works as a heat pump in winter: by the expense of energy of surroundings (refrigerator) it heats the room (heater).

Problem № 3.65

The room is heated by the heat pump. It represents a reversed conditioner,by using of which heat transfer from surroundings to room is occurred. Efficiency of thermal pump is 9.

What is the ratio of heat obtained by the room to work expended on pumping this heat from outside?

Solution:

Efficiency of heat pump is:

$$\eta' = \frac{q_c}{A} \tag{3.19}$$

where q_c is heat taken from surroundings (refrigerator), A represents work expended for taking this heat.

The following relationship exists between heats, taken from surroundings (q_c) and obtained by room (q_h) :

$$q_h = |q_c| + A \tag{3.18}$$

Let us transform expression (3.18):

$$\frac{q_h}{A} = \frac{q_c}{A} + 1$$

According to condition of problem,

$$\eta' = \frac{q_c}{A} = 9$$

Then

$$\frac{q_h}{A} = 9 + 1 = 10$$

Thus, by expending of 1 unit of energy 10–fold more energy is possible to pump from the outside.

Problem № 3.66

The heat engine works within $50 \div 200^0 C$ by using of heat Q, evolved at the fuel combustion. Its real efficiency is equal to $0.5 \times \eta_{theor.}$ This heat engine sets in motion a heat pump.

Calculate the efficiency of a heat pump, if the temperature of surroundings is $7^\circ C$ and temperature of the room equals to $17^\circ C$.

Solution:

The theoretical efficiency of heat engine is equal to the efficiency of Carno's direct cycle (3.17):

$$\eta_{theor.} = \frac{T_h - T_c}{T_h} = \frac{200 - 50}{200 + 273} = 0.317$$

According to the condition of sum,

$$\eta_{real.} = 0.5 \times \eta_{theor.} = 0.5 \times 0.317 = 0.159$$

i.e. only 0.159 fraction (15.9 %) of the heat evolved in the fuel combustion is transformed into the work in the heat engine:

$$A = 0.159 \, Q$$

This work sets in motion the heat pump. The later represents the cooling plant, by which the heat is transfered from the refrigerator (surroundings) to the heater (room).

The upper limit of efficiency of the cooling engine is equal to the efficiency of Carno's reversed cycle:

$$\eta' = \frac{T_c}{T_h - T_c} = \frac{q_c}{A} \qquad (3.19)$$

where T_c is the temperature of refrigerator (surroundings), T_h represents the temperature of heater (room), q_c is heat, taken away from the refrigerator and A represents work, which sets in action the heat pump.

By the condition of problem, $T_c = T_{surr.} = 280K$, $T_h = T_{room} = 290K$.

Then according to (3.19),

$$\eta' = \frac{280}{17 - 7} = 28$$

In the same time,

$$\eta' = \frac{q_c}{A}$$

Hence

$$q_c = \eta' \cdot A = 28 \times 0.159 Q = 4.44 Q.$$

As is seen, by using of the heat pump the room obtains 4.44 Q heat; by the direct combustion of the fuel at the best heat Q would be obtained by the room. Therefore, the efficiency of heat pump as the means of heating is doubtless.

The efficiency of the heat pump is $\eta' = 28$ in our problem. This means, if 1J work is expensed, 28 J of heat may be obtained from surroundings. The heat pump represents an amplifier of energy: on the basis of small amount of work the large amount of heat is obtained. Therefore, the heat pump (the reversed conditioner) is sufficiently widely used in the life already.

Remark: The use of heat pump possesses one more significant aspect besides of efficiency. There is the tremendous resource of energy in surroundings, which is dissipated in very large volume and exists at relatively low temperature. Therefore this energy is faulty in order to obtain work from it. But that, what is useless for work, may be successfully used for heating. By using the heat pump the collecting and amplifying of energy of the "inferior quality" are carried out in heating purposes. The performance of certain work is really necessary for the concentration of energy of low quality, however its amount is insignificant in comparison with heat, obtained from surroundings.

It should be mentioned, that the use of heat pumps on a large scale might induce undesirable ecological changes, viz. reducing of temperatures of the atmosphere and soil, which can cause the unpredictable results. But such danger does not exist yet, due to the comparatively limited use of heat pumps.

Problem № 3.67

The values of molar enthalpy change and temperature at phase transformations of water in standard conditions are following:

$$T_{melt.} = 273.15K; \quad \overline{\Delta H}_{melt.} = 6.009 \, kJ \, / \, mol;$$
$$T_{evap.} = 373.15K; \quad \overline{\Delta H}_{evap.} = 40.7 \quad kJ \, / \, mol.$$

Temperature dependence of heat capacity of H_2O is illustrated in Table 3.7.

Table 3.7. Temperature dependence of heat capaciry of H_2O

T K	C_P, J / K×mol	T K	C_P, J/ K×mol
ice		water	
10	0.13	273.15	(76.08)*
20	1.76	293	75.66
30	4.39	313	75.24
40	6.27	333	75.24
60	10.16	353	75.24
80	13.38	373.15	(75.24)*
100	16.05	vapor	
120	19.23	373.15	(33.84)*
150	21.95	400	34.11
160	22.99	450	34.53
200	27.84	500	35.03
240	33.02	550	35.53
250	35.36		
260	38.50		
270	46.65		
273.15	(49.22)*		

- The values of $\overline{c}_p$ for phase transformations are calculated by linear extrapolation

I

1)Plot the dependence $\overline{C}_P - T$;

2) Determine the values of heat capacity of ice, water and water vapor at temperatures of phase transformation;

3) Determine entropy changes of water at melting and evaporation.

II

1) Plot the dependence $\dfrac{\overline{c}_P}{T} - T$ on the basis of obtained results;

2) Determine entropy changes of ice, water and water vapor in the respective temperature range: 0÷273.15 K;   273.15÷373.15 K and 373.15÷550 K;

3) Calculate the entropy value of water vapor at 550K.

III

1) Determine the absolute value of entropy of H_2O for each temperature given in Table 3.7;

2) Plot the dependence $S—T$.

Solution:

I

1) As is seen from Fig.3.2, which is constructed following data of Table 3.7, heat capacity of ice considerably increases with temperature. The most significant increase takes place in range 260÷270 K or near the melting temperature.

C_P of water exceeds C_P of ice , but in contrast to ice, C_P of water remains practically invariable within the whole range of existence of water in liquid state.

Heat capacity of vapor is much less than heat capacity of water and increases negligibly with temperature.

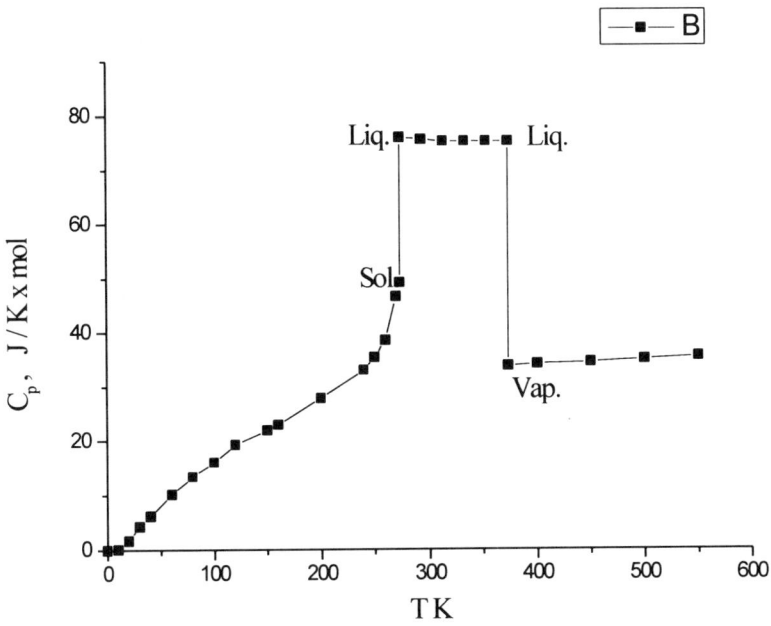

Fig.3.2. Temperature dependence of heat capacity of H_2O

2) The experimental values of heat capacity of water at temperatures of phase transformation are absent in the Table 3.7. Let us determine these values by linear extrapolation.

Ice 273.15 K

The last experimental value of C_P for ice is determined at 270 K (see Table 3.7 and Fig.3.2). Since temperature range 270÷273.15 K is small, it may be considered, that dependence $\overline{C}_P - T$ in this temperature range possesses the same character as in previous one (260÷270 K). Therefore the linear extrapolation is possible:

$$\overline{c}_{p(273.15K)ice} = \overline{c}_{p\,270K} + \frac{\overline{c}_{p\,270K} - \overline{c}_{p\,260K}}{270 - 260}(273.15 - 270) =$$

$$= 46.65 + \frac{46.65 - 38.50}{10} \times 3.15 = 49.22\,J\,/\,K \times mol$$

Water 273.15 K

As is already mentioned, heat capacity of water is invariable in temperature range $273.15 \div 373.15$ K, but a slight decrease of $\overline{C}_P$ takes place in temperature range $293 \div 313$ K (Table 3.7). Let us assume, that this tendency reveals since melting point and carry out the linear extrapolation:

$$\overline{c}_{p(273.15)wat.} = \overline{c}_{p\,293K} + \frac{\overline{c}_{p\,293K} - \overline{c}_{p\,313K}}{313 - 293}(293 - 273.15) =$$

$$= 75.66 + \frac{75.66 - 75.24}{20} \times 19.85 = 76.08\,J\,/\,K \times mol$$

Water 373.15 K

C_P of water is invariable in the temperature range $313 \div 353$ K (see Table 3.7 and Fig.3.2). Therefore it may be assumed, that C_P has a same value at evaporation temperature:

$$\overline{c}_{p(373.15K)wat.} = 75.24\,J\,/\,K \times mol$$

Vapor 373.15 K

Heat capacity of vapor increases linearly in temperature range $400 \div 550$K. Therefore the linear extrapolation is possible until 373.15 K:

$$\overline{C}_{p(373.15K)vap.} = \overline{C}_{p\,400K} - \frac{\overline{C}_{p\,450K} - \overline{C}_{p\,400K}}{450 - 400}(400 - 373.15) =$$

$$= 34.11 - \frac{34.53 - 34.11}{50} \times 26.85 = 33.88\,J\,/\,K \times mol$$

(The values for H_2O obtained by extrapolation are given in Table 3.7 by using of parentheses).

3) In the condition of problem is given, that melting of ice under 1 atm pressure proceeds at constant temperature 273.15 K and evaporation of water at constant 373.15 K. The reason of proceeding of these phase transformations at constant temperature is following: energy delivered to system at melting and evaporation is expended on destruction of existing intermolecular bonds. This process continues till the last molecule transfers into new state of aggregation. If delivery of energy is further continued, the temperature of new state of aggregation will increase. As concerning solidification and condensation, energy released at these phase transformations, is transferred to surroundings in the form of heat. Otherwise, temperature of system becomes higher,than it corresponds to new state of aggregation and phase transformation will not take place.

Let us determine entropy change in processes of ice melting and water evaporation. According to expression (3.25):

$$\Delta S_{ph,transf.} = \frac{\Delta H_{ph.tr.}}{T_{ph.tr.}}$$

It follows that

$$\Delta \overline{S}_{melt.} = \frac{6009}{273.15} = 22\,J\,/\,K \times mol$$

$$\Delta \overline{S}_{evap.} = \frac{40700}{373.15} = 109.1\,J\,/\,K \times mol$$

II

1) The value of $\dfrac{\overline{c_p}}{T}$ in temperature range $10 \div 550$ K was calculated by using of data given in Table 3.7 (see Table 3.8).

Table 3.8. Temperature dependence of c_p / T of H_2O

T K	C_p / T, J/ (K^2×mol)	T K	C_p / T, J/ (K^2×mol)
ice		**water**	
10	0.013	273.15	(0.279)*
20	0.088	293	0.258
30	0.146	313	0.240
40	0.157	333	0.226
60	0.169	353	0.213
80	0.167	373.15	(0.202)*
100	0.161	**vapor**	
120	0.160	373.15	(0.091)*
150	0.146	400	0.085
160	0.144	450	0.077
200	0.139	500	0.070
240	0.138	550	0.065
250	0.141		
260	0.148		
270	0.173		
273.15	(0.180)*		

* The values of C_p / T for phase transformations are obtained from values of C_p, calculated by extrapolation.

Let us plot dependence: $\dfrac{\overline{c_p}}{T} - T$ (Fig.3.3). As is seen from figure, dependence of $\dfrac{\overline{c_p}}{T}$ on temperature is characterized by existing of extremums. As to liquid and vapor states, smooth decrease of $\dfrac{\overline{c_p}}{T}$ takes place with increasing of temperature. The value of $\dfrac{\overline{c}_p}{T}$, (as well as $\overline{c}_p$), abruptly changes in the points of phase transformations.

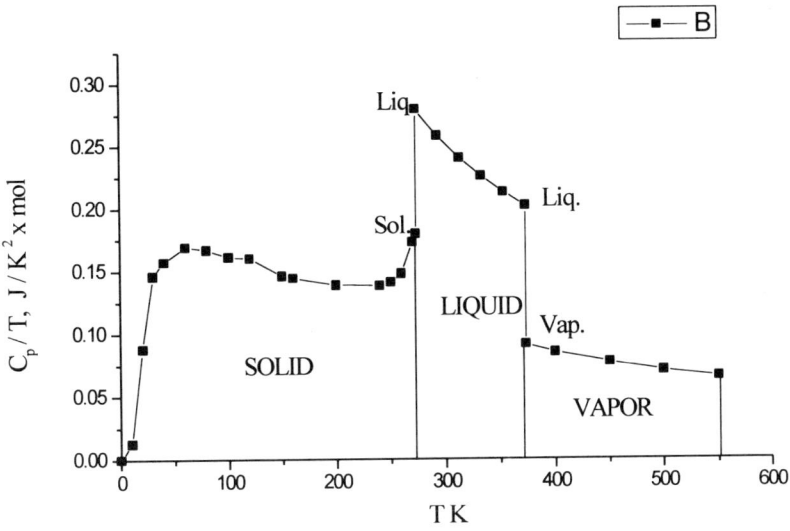

Fig.3.3. Dependence of C_p / T of H_2O on temperature

2) Let us use following expression for determination of entropy change of ice, water and vapor at the change of temperature:

$$\Delta \overline{S} = \int_{T_1}^{T_2} \frac{\overline{c}_p}{T} dT \qquad (3.28)$$

This integral may be calculated by graphic manner. The value of expression (3.28) is area, circumscribed from above by curve of dependence

$\dfrac{\overline{c}_p}{T} - T$, from below by temperature axis and from lateral sides by values

of $\dfrac{\overline{c}_{p_1}}{T_1}$ and $\dfrac{\overline{c}_{p_2}}{T_2}$.

Entropy is an additive quantity. Consequently, ΔS of each state of aggregation may be represented as sum of several ΔS (or of respective areas). The dependence $\dfrac{\overline{c}_p}{T} - T$ has a distinct character for ice, water and water vapor (Fig.3.3). Therefore, areas, corresponded to them, will be calculated by different ways.

Ice 0÷273.15 K

0÷10 K

The experimental data about dependence of heat capacity on temperature at very low temperatures is scare, however it is known, that this dependence is described by Debye's law of T ^{3} :

$$c_p = \alpha T^3 \tag{a}$$

where α is a constant, which characterizes a substance.

Let us introduce expression (a) into (3.28):

$$\Delta \overline{S} = \int_{T_1}^{T_2} \frac{\overline{c}_p}{T} dT = \int_{T_1}^{T_2} \frac{\alpha T^3}{T} dT = \alpha \int_{T_1}^{T_2} T^2 dT \tag{b}$$

$\overline{c}_p$ of water at 10 K is equal to 0.13 J /K×mol according to Table 3.7. Thus,

$$\overline{c}_p = \alpha T^3 = 0.13$$

It follows that

$$\alpha = \frac{0.13}{T^3} = \frac{0.13}{10^3} = 0.13 \times 10^{-3}$$

If this value of α is introduced into expression (b), it is obtained:

$$\overline{\Delta S} = \alpha \int_0^{10} T^2 dT = 0.13 \times 10^{-3} \times \frac{10^3}{3} = \frac{0.13}{3} = 0.043 \, J / K \times mol$$

Remark: Increment of entropy at very low temperatures by change of temperature from 0 to T K is equal to one third of C_P, corresponded to temperature T:

$$\Delta S_{0 \div TK} = S_T = \frac{c_{p(T)}}{3}$$

Let us calculate the value of ΔS of ice above 10 K by a graphic method.

Area of figure corresponded to solid state may be represented by sum of areas of several trapeziums. Because of existence of extremums on the curve $\frac{\overline{c_p}}{T} - T$ of ice, the area of the net figure will be divided into parts according to character of curve. (In the given case dividing into the equal parts is impossible).

Area of trapezium is expressed so:

$$\Omega_{trap.} = \frac{y_1 + y_2}{2} \times h \tag{c}$$

where y_1 and y_2 are lengthes of sides of trapezium and h represents width of base.

$10 \div 30$ K

$$\Delta \overline{S}_{(10 \div 30K)} = \int_{10}^{30} \frac{\overline{c}_p}{T} dT = \Omega_{trap.} = \frac{0.013 + 0.146}{2}(30 - 10) = 1.590 \, J \,/\, K \times mol$$

$30 \div 60$ K

$$\Delta \overline{S}_{(30 \div 60K)} = \frac{0.146 + 0.169}{2}(60 - 30) = 4.725 \, J \,/\, K \times mol$$

$60 \div 160$ K

$$\Delta \overline{S}_{(60 \div 160K)} = \frac{0.169 + 0.144}{2}(160 - 60) = 15.650 \, J \,/\, K \times mol$$

$160 \div 240$ K

$$\Delta \overline{S}_{(160 \div 240K)} = \frac{0.144 + 0.138}{2}(240 - 160) = 11.280 \, J \,/\, K \times mol$$

$240 \div 260$ K

$$\Delta \overline{S}_{(240 \div 260K)} = \frac{0.138 + 0.148}{2}(260 - 240) = 2.860 \, J \,/\, K \times mol$$

$260 \div 273.15$ K

$$\Delta \overline{S}_{(260 \div 273.15K)} = \frac{0.148 + 0.180}{2}(273.15 - 260) = 2.157 \, J \,/\, K \times mol$$

$$\Delta \overline{S}_{(0 \div 273.15K)ice} = \Delta \overline{S}_{(0 \div 10K)} + \Delta \overline{S}_{(10 \div 30K)} + \Delta \overline{S}_{(30 \div 60K)} + \Delta \overline{S}_{(60 \div 160K)} +$$

$$+ \Delta \overline{S}_{(160 \div 240K)} + \Delta \overline{S}_{(240 \div 260K)} + \Delta \overline{S}_{(260 \div 273.15K)} = 0.043 + 1.590 + 4.725 +$$

$$+ 15.650 + 11.280 + 2.860 + 2.157 = 38.305 \, J \,/\, K \times mol$$

Water, 273.15÷373.15 K

In contrast to ice, dependence $\dfrac{\bar{c}_p}{T} - T$ for water possesses with a simple character (see Fig.3.3). Therefore the classical mode of method of trapeziums may be used for calculation of ΔS corresponded to liquid state: let us divide the respective temperature range into the equal parts, draw lines from T–axis to crossing the curve, connect their apexes and consider the sum of areas of obtained trapeziums as area of figure corresponded to liquid state. Then

$$\Delta \bar{S}_{(273.15÷373.15K)} = \Omega_{wat.} = \left(\frac{1}{2} y_1 + y_2 + ... + \frac{1}{2} y_k \right) \times h \qquad (d)$$

where h is the base of trapezium and y_1, $y_2,...y_k$ represent sides of respective trapeziums.

Assume, that $h = \Delta T = 20^0$. Then $y_1 = 0.279$; $y_2 = 0.258$; $y_3 = 0.240$;

$y_4 = 0.226$; $y_5 = 0.213$; $y_6 = 0.202$ (see Table 3.8).

According to expression (d):

$$\Delta \bar{S}_{(273.15÷373.15K)wat.} = \Omega_{wat.} =$$

$$= \left(\frac{1}{2} \times 0.279 + 0.258. + 0.240 + 0.226 + 0.213 + \frac{1}{2} \times 0.202 \right) \times 20 =$$

$$= 23.550 \ J/K \times mol$$

Vapor, 373.15÷550 K

Dependence $\dfrac{\bar{c}_p}{T} - T$ for vapor is practically linear (Fig.3.3). Therefore the figure corresponded to liquid state may be considered as trapezium. Then according expression (c),

$$\Delta \overline{S}_{(373.15 \div 550K)} = \Omega_{vap.} = \frac{y_1 + y_2}{2} \times h = \frac{0.091 + 0.065}{2}(550 - 373.15) =$$
$$= 13.794 \ J / K \times mol$$

3) Let us calculate the value of molar entropy of water vapor at 550 K. In accordance with Planck's postulate, entropy of pure cristalline substance at absolute zero is equal to zero:

$$S_0 = 0 \qquad\qquad (3.38)$$

Hence,

$$\Delta S_{(0 \div TK)} = S_{TK} - S_0 = S_T \qquad\qquad (e)$$

On the other hand, entropy is the additive quantity and when $T > T_{boil.}$,

$$\Delta S_{(0 \div TK)} = \Delta S_{(0 \div T_{melt.})} + \Delta S_{melt.} + \Delta S_{(T_{melt.} \div T_{boil.})} + \Delta S_{boil.} + \Delta S_{(T_{boil.} \div T)} \qquad (f)$$

Let us introduce results obtained by us into expression (f):

$$\Delta \overline{S}_{(0 \div 550K)} = 38.305 + 22 + 23.550 + 109.1 + 13.794 = 206.749 \ J / K \times mol$$

According expression (e),

$$\Delta \overline{S}_{(0 \div 550K)} = \overline{S}_{550K} = 206.749 \ J / K \times mol$$

III

Let us determine the absolute value of entropy of H_2O at each temperature, given in Table 3.7.

Ice

10 K

As is already calculated, $\Delta \overline{S}_{(0 \div 10K)} = \overline{S}_{10K} = 0.043 \, J/K \times mol$

The method of trapezium is used for determination of entropy values at temperatures, which exceed 10 K.

20 K

$$\Delta \overline{S}_{(10 \div 20K)} = \Omega_{trap.(10 \div 20K)} = \frac{y_1 + y_2}{2} \times h = \frac{0.013 + 0.088}{2}(20 - 10) =$$

$$= 0.505 \, J/k \times mol$$

$$\overline{S}_{20K} = \Delta \overline{S}_{(10 \div 20K)} + \overline{S}_{10K} = 0.505 + 0.043 = 0.548 \, J/K \times mol$$

30 K

$$\Delta \overline{S}_{(20 \div 30K)} = \frac{0.088 + 0.146}{2}(30 - 20) = 1.170 \, J/K \times mol$$

$$\overline{S}_{30K} = 1.170 + 0.548 = 1.718 \, J/K \times mol$$

40 K

$$\Delta \overline{S}_{(30 \div 40K)} = \frac{0.146 + 0.157}{2}(40 - 30) = 1.515 \, J/K \times mol$$

$$\overline{S}_{40K} = 1.515 + 1.718 = 3.233 \, J/K \times mol$$

60 K

$$\Delta \overline{S}_{(40 \div 60K)} = \frac{0.157 + 0.169}{2}(60 - 40) = 3.260 \, J/K \times mol$$

$$\overline{S}_{60K} = 3.260 + 3.233 = 6.493 \, J/K \times mol$$

80 K

$$\Delta \overline{S}_{(60 \div 80K)} = \frac{0.169 + 0.167}{2}(80 - 60) = 3.360 \, J/K \times mol$$

$$\overline{S}_{80K} = 3.360 + 6.493 = 9.853 \, J/K \times mol$$

100 K

$$\Delta \bar{S}_{80 \div 100K} = \frac{0.167 + 0.161}{2} = (100 - 80) = 3.280 \, J \, / \, K \times mol$$

$$\bar{S}_{100K} = 3.280 + 9.853 = 13.133 \, J \, / \, K \times mol$$

120 K

$$\Delta \bar{S}_{100 \div 120K} = \frac{0.161 + 0.160}{2} (120 - 100) = 3.210 \, J \, / \, K \times mol$$

$$\bar{S}_{120K} = 3.210 + 13.133 = 16.343 \, J \, / \, K \times mol$$

150 K

$$\Delta \bar{S}_{(120 \div 150K)} = \frac{0.160 + 0.146}{2} (150 - 120) = 4.590 \, J \, / \, K \times mol$$

$$\bar{S}_{150K} = 4.590 + 16.343 = 20.933 \, J \, / \, K \times mol$$

160 K

$$\Delta \bar{S}_{(150 \div 160K)} = \frac{0.146 + 0.144}{2} (160 - 150) = 1.450 \, J \, / \, K \times mol$$

$$\bar{S}_{160K} = 1.450 + 20.933 = 22.383 \, J \, / \, K \times mol$$

200 K

$$\Delta \bar{S}_{(160 \div 200K)} = \frac{0.144 + 0.139}{2} (200 - 160) = 5.660 \, J \, / \, K \times mol$$

$$\bar{S}_{200K} = 5.660 + 22.383 = 28.043 \, J \, / \, K \times mol$$

240 K

$$\Delta \bar{S}_{(200 \div 240K)} = \frac{0.139 + 0.138}{2} (240 - 200) = 5.540 \, J \, / \, K \times mol$$

$$\bar{S}_{240K} = 5.540 + 28.043 = 33.583 \, J \, / \, K \times mol$$

250 K

$$\Delta \overline{S}_{(240 \div 250K)} = \frac{0.138 + 0.141}{2}(250 - 240) = 1.395 \, J \, / \, K \times mol$$

$$\overline{S}_{250K} = 1.395 + 33.583 = 34.978 \, J \, / \, K \times mol$$

260 K

$$\Delta \overline{S}_{(250 \div 260K)} = \frac{0.141 + 0.148}{2}(260 - 250) = 1.445 \, J \, / \, K \times mol$$

$$\overline{S}_{260K} = 1.445 + 34.978 = 36.423 \, J \, / \, K \times mol$$

270 K

$$\Delta \overline{S}_{(260 \div 270K)} = \frac{0.148 + 0.173}{2}(270 - 260) = 1.605 \, J \, / \, K \times mol$$

$$\overline{S}_{270K} = 1.605 + 36.423 = 38.028 \, J \, / \, K \times mol$$

273.15 K

$$\Delta \overline{S}_{(270 \div 273.15K)} = \frac{0.173 + 0.180}{2}(273.15 - 270) = 0.556 \, J \, / \, K \times mol$$

$$\overline{S}_{273.15K \, (ice)} = 0.556 + 38.028 = 38.584 \, J \, / \, K \times mol$$

Water

273.15 K

We have already calculated, that entropy change at melting of ice under 1 atm pressure is:

$$\Delta \overline{S}_{melt.} = \frac{\Delta \overline{H}_{melt.}}{T_{melt.}} = \frac{6009}{273.15} = 22 \, J \, / \, K \times mol$$

On the other hand,

$$\Delta \overline{S}_{melt.} = \overline{S}_{(273.15K)wat.} - \overline{S}_{(273.15K)ice}$$

It follows that

$$\overline{S}_{(273.15K)wat.} - \overline{S}_{(273.15K)ice} = 22 \, J \,/\, K \times mol$$

and

$$\overline{S}_{(273.15K)wat.} = 22 + \overline{S}_{(273.15K)ice} = 22 + 38.584 = 60.584 \, J \,/\, K \times mol$$

293 K

$$\Delta \overline{S}_{(273.15 \div 293K)} = \frac{0.279 + 0.258}{2}(293 - 273.15) = 5.330 \, J \,/\, K \times mol$$

$$\overline{S}_{293K} = 5.330 + 60.584 = 65.914 \, J \,/\, K \times mol$$

313 K

$$\Delta \overline{S}_{(293 \div 313K)} = \frac{0.258 + 0.240}{2}(313 - 293) = 4.980 \, J \,/\, K \times mol$$

$$\overline{S}_{313K} = 4.980 + 65.914 = 70.894 \, J \,/\, K \times mol$$

333 K

$$\Delta \overline{S}_{(313 \div 333K)} = \frac{0.240 + 0.226}{2}(333 - 313) = 4.660 \, J \,/\, K \times mol$$

$$\overline{S}_{333K} = 4.660 + 70.894 = 75.554 \, J \,/\, K \times mol$$

353 K

$$\Delta \overline{S}_{(333 \div 353K)} = \frac{0.226 + 0.213}{2}(353 - 333) = 4.390 \, J \,/\, K \times mol$$

$$\overline{S}_{353K} = 4.390 + 75.554 = 79.944 \, J \,/\, K \times mol$$

373.15 K

$$\Delta \overline{S}_{(353 \div 373.15K)} = \frac{0.213 + 0.202}{2}(373.15 - 353) = 4.181 J / K \times mol$$

$$\overline{S}_{(373.15K)wat.} = 4.181 + 79.944 = 84.125 J / K \times mol$$

Vapor

373.15 K

Entropy change at evaporation of water is:

$$\Delta \overline{S}_{evap.} = \frac{\Delta \overline{H}_{evap.}}{T_{evap.}} = \frac{40700}{373.15} = 109.1 J / K \times mol$$

In addition to:

$$\Delta \overline{S}_{evap.} = \overline{S}_{(373.15K)vap.} - \overline{S}_{(373.15K)wat.}$$

Accordingly,

$$\overline{S}_{(373.15K)vap.} - \overline{S}_{(373.15K)wat.} = 109.1 J / K \times mol$$

and

$$\overline{S}_{(373.15K)vap.} = 109.100 + 84.110 = 193.210 J / K \times mol$$

400 K

$$\Delta \overline{S}_{(373.15 \div 400K)} = \frac{0.091 + 0.085}{2}(400 - 373.15) = 2.363\,J\,/\,K \times mol$$

$$\overline{S}_{400K} = 2.363 + 193.21 = 195.573\,J\,/\,K \times mol$$

450 K

$$\Delta \overline{S}_{(400 \div 450K)} = \frac{0.085 + 0.077}{2}(450 - 400) = 4.05\,J\,/\,K \times mol$$

$$\overline{S}_{450K} = 4.050 + 195.573 = 199.623\,J\,/\,K \times mol$$

500 K

$$\Delta \overline{S}_{(450 \div 500K)} = \frac{0.077 + 0.070}{2}(500 - 450) = 3.675\,J\,/\,K \times mol$$

$$\overline{S}_{500K} = 3.675 + 199.623 = 203.298\,J\,/\,K \times mol$$

550 K

$$\Delta \overline{S}_{(500 \div 550K)} = \frac{0.070 + 0.065}{2}(550 - 500) = 3.375\,J\,/\,K \times mol$$

$$\overline{S}_{550K} = 3.375 + 203.298 = 206.673\,J\,/\,K \times mol$$

2) Let us tabulate the obtained values of entropy (Table3.9) and plot dependence $S{-}T$ (Fig.3.4). As is seen from Table 3.9 and Fig.3.4, entropy increases with temperature in all three states of aggregation. A significant increase of entropy takes place at phase transformations, especially, at evaporation. The reason of this is conditioned by following: entropy represents a measure of disorder. Disorder increases with temperature and it is revealed in entropy increase of all state of aggregation. An abrupt change of entropy at phase transformations is caused by different degree of ordering of substance in various state of aggregation e.g. liquid state is less disordered than solid one, but vapor state is much greater disordered, than liquid one. Just that is why

$$S_{vapor} > S_{liquid} > S_{solid}$$

and

$$\Delta S_{evap.} \gg \Delta S_{melt.}$$

Table 3.9. Entropy of H_2O in temperature range $0 \div 550$ K.

(P = 1 atm)

T K	S, J / K × mol	T K	S J / K × mol
ice		**water**	
0	0	273.15	60.58
10	0.04	293	65.91
20	0.55	313	70.89
30	1.72	333	75.55
40	3.23	353	79.94
60	6.49	373.15	84.13
80	9.85	**vapor**	
100	13.13	373.15	193.21
120	16.34	400	195.57
150	20.93	450	199.62
160	22.38	500	203.30
200	28.04	550	206.67
240	33.58		
250	34.98		
260	36.42		
270	38.03		
273.15	38.58		

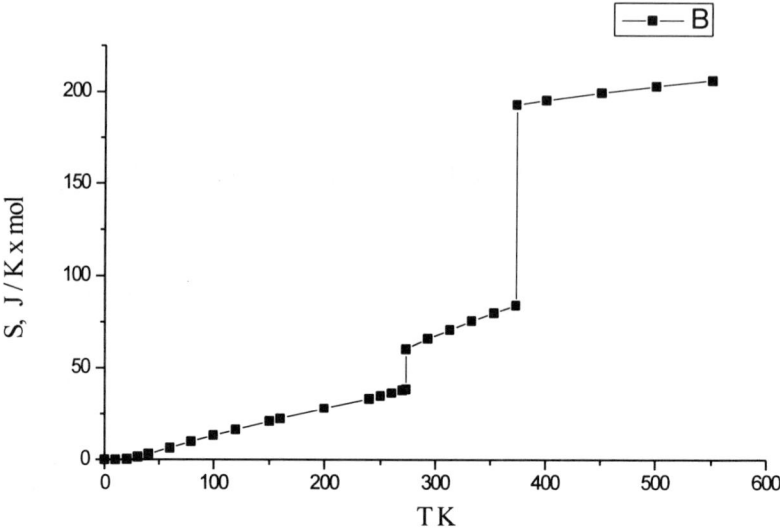

Fig.3.4. Dependence of entropy of H_2O on temperature

4. Thermodynamic Potentials

Theoretical Part

Characterization of process by entropy change of "universe" is related with many difficulties. The difficulty in the determination of entropy change of surroundings represents one of them. Therefore the values, characterizing the system only, so–called thermodynamic potentials are introduced in thermodynamics. By using of them determination of equilibrium state and spontaneity of process is possible. Thermodynamic potentials are:

$F = U - TS$ (Helmholtz's function)

$G = H - TS$ (Gibbs's function)

$\Delta F = \Delta U - T\Delta S \leq 0$ (V,T = const)   (4.1)

$\Delta G = \Delta H - T\Delta S \leq 0$ (P,T = const)   (4.2)

At equilibrium:

$\Delta F = 0$ (V,T = const)

$\Delta G = 0$ (P,T = const)

In the spontaneous process:

$\Delta F < 0$ (V,T = const)

$\Delta G < 0$ (P,T = const)

243

ΔF and ΔG are related with work as follows:

$$\Delta F = \Delta U - T\Delta S \leq A' \qquad \text{(V,T = const)} \qquad (4.3)$$

$$\Delta G = \Delta H - T\Delta S \leq A' \qquad \text{(P,T = const)} \qquad (4.4)$$

where A' is so–called useful work.

In expressions (4.3) and (4.4) designations "=" and "<" correspond to reversible and irreversible processes respectively. More work is fulfilled in reversible process than in irreversible one (according to our designations, work fulfilled by system: $A' < 0$).

F and G are state functions. Therefore, their changes must be identical in both reversible and irreversible processes, if the initial and final stages are the same.

ΔF and ΔG, besides equations (4.3) and (4.4) may be expressed as follows:

$$\Delta F = -T\Delta_i S + A' \qquad \text{(V,T = const)} \qquad (4.5)$$

$$\Delta G = -T\Delta_i S + A' \qquad \text{(P,T = const)} \qquad (4.6)$$

where $T\Delta_i S$ represents energy, which is stipulated by entropy "arising" (formation) in the system. It is equal to Clausius's "noncompensated heat" (q') and to so–called "lost work" (A'_{lost}):

$$T\Delta_i S = q' = A'_{lost} \qquad (4.7)$$

(see Chapter 3, p.116).

Entropy is not arisen in the reversible process:

$$T\Delta_i S = q' = A'_{lost} = 0$$

and change of thermodynamic potential of system is wholly expended on the work:

$$\Delta G = A'_{max} \qquad (4.8)$$

Change of thermodynamic potential (ΔG) in irreversible pocesses is distributed between work (A') and entropy formation ($T\Delta_i S$). Therefore the system fulfils less work, than in reversible process, but entropy increases more (or decreases less), than it is corresponded to heat received (or released) by system. Despite the difference in ways of redistribution of potential energy of system, its amount obeying these redistributions is the same: $\Delta G_{revers.} = \Delta G_{irrev.}$

So–called characterizing functions are frequently used in thermodynamics. If total description of state of system or establishing of numerical values of thermodynamic quantities are possible via any function and variables related with it, the function is characterizing. The characterizing functions are: U *(S, V)*; H *(S, P)*; S *(U, V)*; F *(T, V)*; G *(T, P)*.

G *(T ,P)* may be written in unfolded form as follows:

$$dG = -SdT + VdP \qquad (4.9)$$

This expression is called as Gibbs's fundamental equation. From (4.9) it is obtained:

$$V = \left(\frac{\partial G}{\partial P} \right)_T \qquad (4.10)$$

$$-S = \left(\frac{\partial G}{\partial T} \right)_p \qquad (4.11)$$

These last expressions correspond to processes, where only change of temperature or pressure takes place. V and S are represented here as change of G, conditioned by change of P or T.

If other process takes place in the system e.g. phase transformation or chemical reaction, then

$$\Delta V = \left(\frac{\partial \Delta G}{\partial P} \right)_T \qquad (4.12)$$

$$-\Delta S = \left(\frac{\partial \Delta G}{\partial T}\right)_P \tag{4.13}$$

where ΔV, ΔS and ΔG represent the difference in V, S and G of substances participating in the process (reactants and products):

$$\Delta V = \sum V_{prod.} - \sum V_{react.} \tag{4.14}$$

$$\Delta S = \sum S_{prod.} - \sum S_{react.} \tag{4.15}$$

$$\Delta G = \sum G_{prod.} - \sum G_{react.} \tag{4.16}$$

but $\left(\dfrac{\partial \Delta G}{\partial P}\right)_T$ and $\left(\dfrac{\partial \Delta G}{\partial T}\right)_P$ are changes of ΔG of chemical reaction (or phase transformation), conditioned by changes of P and T.

Combination of expressions (4.2) and (4.13) gives:

$$\Delta G = \Delta H + T\left(\frac{\partial \Delta G}{\partial T}\right)_P \tag{4.17}$$

This equation is called Gibbs–Helmholtz's equation. Let us introduce in this equation following expressions:

$$\Delta G = A'_{max} \tag{4.8}$$

and

$$\Delta H = Q_p \tag{2.4}$$

where A'_{max} is maximal useful work of reversible process, but Q_p represents heat of irreversible process. Then it is obtained:

$$A'_{\max} = Q_p + T\left(\frac{\partial A'_{\max}}{\partial T}\right)_P \tag{4.18}$$

Expression (4.18) represents Gibbs–Helmholtz's equation also, but it contains quantities, determination of which is possible via experiments, viz. heat (Q_p), maximal useful work (A'_{max}) and temperature coefficient of this work $\left(\dfrac{\partial A'_{\max}}{\partial T}\right)$.

Problems

Problem № 4.1

The values of molar enthalpy and molar entropy change of glucose oxidation reaction

$$C_6H_{12}O_6 + 6O_2 \rightarrow 6CO_2 + 6H_2O$$

at standard conditions are equal: $\Delta\overline{H} = -2808 \times 10^3$ J / mol, $\Delta\overline{S} = 182.4$ J / mol×K.

1) Does the reaction proceed spontaneously at standard conditions (1 atm, 25^0C)?

2) What is A'_{max} of reversible process?

3) What are entropy changes of surroundings and "universe" in reversible process?

Solution:

1) In order to establish the spontaneity of process let us refer to expression (4.2):

$$\Delta G = \Delta H - T\Delta S = -2808 \times 10^3 - 298 \times 182.4 = -2862.355 \times 10^3 \, J \, / \, mol$$

$\Delta G < 0$, i.e. oxidation of glucose in standard conditions proceeds spontaneously.

2) In reversible process at $P,T = const$ conditions

$$A'_{max} = \Delta G = -2862.355 \times 10^3 \, J \, / \, mol$$

Work, which may be fulfilled due to this reaction, exceeds the energy, released in the system ($A'_{max} = -2862 \times 10^3$, but $\Delta \overline{H} = -2808 \times 10^3$ J / mol). The reason of this paradoxical result will be given below.

3) For reversible proceeding of the process following condition must be fulfilled:

$$\Delta S_{"univ"} = \Delta S_{syst.} + \Delta S_{surr.} = 0 \qquad (3.46)$$

It follows:

$$\Delta S_{surr.} = -\Delta S_{syst.} = -182.4 J \, / \, K \times mol$$

Entropy of surroundigs changes only due to heat exchange with the system:

$$\Delta S_{surr.} = \frac{q_{surr.}}{T} \qquad (3.44)$$

Hence

$$q_{surr.} = T\Delta S_{surr.} = 298 \times (-182.4) = -54.355 \times 10^3 \, J \, / \, mol$$

Thus, surroundings transfers to system 54.355×10^3 *J/mol* heat.

At the same time, $\Delta H = -2808 \times 10^3$ *J/mol* energy is released in the system as a result of proceeding chemical reaction. The system can not transfer this energy to surroundings, because surroundings transfers heat to system vice versa (otherwise the condition of reversibility: $T\Delta S_{"univ."} = 0$ will be not fulfilled). Therefore the system is obliged to transform into work both energy released in the system (ΔH) and heat received from surroundings ($q_{surr.}$):

$$A'_{max} = \Delta H + q_{surr.} = \Delta H - q_{syst.} = \Delta H - T\Delta S =$$
$$= -2808 \times 10^3 - 54.355 \times 10^3 = -2862.355 \times 10^3 \ J \ / \ mol$$

Because of this reason: $|A'_{max}| > |\Delta H|$. All processes proceeded by enthalpy decrease $(\Delta H < 0)$ and entropy increase $(\Delta S > 0)$ are characterized with such relationship between A'_{max} and ΔH.

Remark: Mammals receive energy necessary for vital functions as a result of slow oxidation of glucose. It is interesting, that in the nature just that process is selected as a source of work, which releases a great deal of energy and moreover makes possible to pump out energy from surroundings.

Problem № 4.2

Oxidation reaction of methane:

$$CH_4 + \tfrac{1}{2} O_2 \rightarrow CH_3OH_{(liquid)}$$

proceeds under conditions of constant pressure and temperature $(P = 1 \ atm, T = 298 \ K)$.

1) What are the values of molar ΔS, ΔH, ΔG of methane oxidation?

2) What relationship must be existed between $T\Delta S$ and ΔH for spontaneous proceeding of reaction?

3) What are equal A' and q_p of reversible process?

4) What are entropy changes of surroundings and "universe" in reversible process?

Solution:

1) ΔS and ΔH we calculate by using of handbook:

$$\Delta S_{syst.} = S_{CH_3OH(liq.)} - S_{CH_4} - \frac{1}{2}S_{O_2} = 126.7 - 186.1 - \frac{1}{2} \times 205.1 =$$
$$= -161.95 J \ / \ K \times mol$$

$$\Delta H_{syst.} = \left(\Delta H_{CH_3OH_{liq.}}\right)_{formation} - \left(\Delta H_{CH_4}\right)_{form.} = -239 \times 10^3 -$$

$$-(-74.81 \times 10^3) = -164.19 \times 10^3 \, J/mol$$

$$\Delta G = \Delta H - T\Delta S = -164.19 \times 10^3 - (-161.95 \times 298) = -115.93 \times 10^3 \, J/mol$$

2) Thus, despite $\Delta S_{syst.} < 0$, oxidation of methane proceeds spontaneously, which is pointed out by $\Delta G_{syst.} < 0$. For spontaneous proceeding of all processes, where $\Delta S_{syst.} < 0$, and $\Delta H_{syst.} < 0$ the condition $|T\Delta S| < |\Delta H|$ must be fulfilled. Otherwise $\Delta G = (\Delta H - T\Delta S) > 0$ and process will not proceed.

3) According to (4.8), at reversible performance of process

$$A'_{max} = \Delta G = -115.93 \times 10^3 \, J/mol$$

Work is fulfilled at the expense of energy release in system $(\Delta H_{syst.} < 0)$ The other part of this energy is transferred to surroundings in the form of heat:

$$q_{syst.}^{revers.} = \Delta H - A'_{max} = \Delta H - \Delta G = T\Delta S = 298 \times (-161.95) =$$

$$= -48.26 \times 10^3 \, J/mol$$

4) In reversible process

$$\Delta S_{"univ."} = \Delta S_{syst.} + \Delta S_{surr.} = 0$$

$$\Delta S_{surr.} = -\Delta S_{syst.} = -(-161.95) = 161.95 J/K \times mol$$

Entropy change of surroundings may be calculated also as follows:

$$\Delta S_{surr.} = \frac{q_{surr.}}{T} = -\frac{q_{syst.}}{T} = -\frac{-48.26 \times 10^3}{298} = 161.95 J / K \times mol$$

Problem № 4.3

Molar values of ΔH and ΔS by dissolving of NH_4Cl in water in standard conditions (P = 1 atm, T = 298 K) are equal to 34.7×10^3 J/mol and 167.1 J / K×mol respectively.

1) Does the solution of NH_4Cl proceed spontaneously under these conditions or not?

2) What are equal heat and A'_{max} of reversible process?

3) How does entropy of surroundings and "universe"change in reversible process?

Solution:

1) $\Delta G = \Delta H - T\Delta S = 34.7 \times 10^3 - 298 \times 167.1 = -15.096 \times 10^3 J / mol$

$\Delta G < 0$ and therefore, despite the endothermic character of process, solution proceeds spontaneously.

2) $$A'_{max} = \Delta G = -15.096 kJ / mol$$

$$q_{syst.}^{revers.} = T\Delta S_{syst.} = 298 \times 167.1 = 49.796 kJ / mol$$

3) $$\Delta S_{surr.} = \frac{q_{surr.}}{T} = -\frac{q_{syst.}}{T} = -\frac{49796}{298} = -167.1 J / K \times mol$$

$$\Delta S_{"univ."} = \Delta S_{syst.} + \Delta S_{surr.} = 167.1 - 167.1 = 0$$

Endothermic process proceeds spontaneously in case, when $T\Delta S > \Delta H$. If the process is characterized by high endothermality or (and) the value of $T\Delta S$ is minor, then $T\Delta S < \Delta H$, $\Delta G > 0$ and the process will not proceed spontaneously. High temperature favors the spontaneity of endothermic process.

Let us tabulate results of problems №№ 4.1÷4.3 (Table 4.1).

As is shown from Table 4.1, entropy of system increases in some reversible process and decreases in some one. With that $\Delta S_{"univ."} = 0$ in all cases. Thus, establishment of spontaneous proceeding of irreversible process is impossible via entropy change of reversible process. The value of ΔG is used for this purpose. $\Delta G < 0$ for reversible process means, that irreversible process proceeds spontaneously, but $\Delta G > 0$ denotes, that process does not run spontaneously.

Table 4.1.The values of thermodynamic functions of some reversible processes

298 K, 1 atm

Process	ΔS_{syst} J/K× ×mol	ΔS_{srr} J/K× ×mol	ΔS_{univ} J/K× ×mol	$T\Delta S_{sys.}$ kJ/mol	$\Delta H_{syst.}$ kJ/mol	$\Delta G = A'_{max}$ kJ/mol	Condition for spontaneous proceeding of process
$C_6H_{12}O_6+$ $+6O_2 \rightarrow$ $\rightarrow 6CO_2+$ $+6H_2O$	182	−182	0	54.4	−2808	−2862	For any ratio of $T\Delta S$ and ΔH
CH_4+ $+0.5\ O_2 \rightarrow$ $\rightarrow CH_3OH_{liq.}$	−162	162	0	−48.3	−164	−116	$\lvert T\Delta S \rvert < \lvert \Delta H \rvert$
Dissolving of NH_4Cl in water	167	−167	0	49.8	34.7	−15	$T\Delta S > \Delta H$

Problem № 4.4

Does diamond transform spontaneously into graphite under conditions 1 atm pressure and room temperature or not?

Solution:

Let us consider allotropic transformation:

$$\text{diamond} \rightarrow \text{graphite}$$

ΔH and ΔS of process we calculate by using of handbook:

$$\Delta H = \left(\Delta H_{graph}\right)_{form.} - \left(\Delta H_{diam.}\right)_{form.} = 0 - 1897 = -1897 J / mol$$

$$\Delta S = S_{graph.} - S_{diam.} = 5.74 - 2.38 = 3.36 J / K \times mol$$

$$\Delta G = \Delta H - T\Delta S = -1897 - 298 \times 3.36 = -2898.28 J / mol$$

$\Delta G = -2898.28$ *J/mol* and therefore the process of transformation of diamond into graphite at standard conditions proceeds spontaneously. The question may be arisen here: if coalification of diamond proceeds spontaneously even in conventional conditions, why we do not observe this or why is considered diamond as a precious stone? The fact is that equilibrium thermodynamics offers the possibility of proceeding of process, but not its real proceeding in time (i.e. rate). The coalification of diamond proceeds actually (as is shown by $\Delta G < 0$), but the rate of this process is so negligible, that we do not observe it and diamond is practically invariable during centuries.

Problem № 4.5
What is a change of Gibbs's function for 1 L ethanol, if pressure increases from 1 to 50 atm at room temperature (assume that ethanol does not compress)?

Solution:
Let us use Gibbs's fundamental equation (4.9):

$$dG = -SdT + VdP$$

Since $T = \text{const}$,
$$dG = VdP$$

According to condition of problem, ethanol does not compress i.e. $V = const$. Then

$$\int_1^2 dG = V \int_1^2 dP$$

It follows that

$$\Delta G = G_2 - G_1 = V\Delta P = V(P_2 - P_1) = 1(50 - 1) = 49\ L \times atm =$$

$$= 49 \times 1.01 \times 10^2\ J = 4949\ J = 4.949\ kJ$$

Problem № 4.6

Determine change of isobaric potential at isothermal compression of 1 m^3 oxygen from 1 to 50 atm (assume, that O_2 is an ideal gas).

Solution:

It is obtained from Gibbs's fundamental equation (4.9) at $T = const$ conditions:

$$dG = VdP$$

$$\int_1^2 dG = \int_1^2 V(P)dP \qquad \text{(a)}$$

The volume of gases strongly depends on pressure. Therefore, in contrast to problem № 4.5, in the given problem $V \neq const$ and volume is a function of pressure. Let us introduce the value of $V = f(P)$ from ideal gas equation : $V = \dfrac{nRT}{P}$ into expression (a). Then it is obtained:

$$\int_1^2 dG = \int_1^2 \frac{nRT}{P}dP = nRT \int_1^2 \frac{dP}{P} = nRT \ln \frac{P_2}{P_1}$$

$$\Delta G = nRT \ln \frac{P_2}{P_1}$$

Let us determine the number of oxygen moles (n).

1 mole of O_2 occupies at 298 K: $22.4 \times \dfrac{298}{273} = 24.45L$.

$1\ m^3\ O_2 = 10^3\ L\ O_2$, which contains $\dfrac{10^3}{24.45} = 40.9\ moles.$

$\Delta G = 40.9 \times 8.31 \times 298 \ln (50/1) = 396225.1\ J = 396,225\ kJ.$

Problem № 4.7

The reaction is given:

$$C_2H_{2\ (g)} + 2\ H_2O_{\ (liq.)} \rightarrow CH_3COOH_{\ (liq.)} + H_{2\ (g)}$$

Standard molar heat of this reaction $\Delta H^0 = -\ 139.97$ kJ/mol; heat by carrying out the process equilibriumly (reversibly) is $q_p = -\ 14.995$ kJ/mol.

What amount of A'_{max} may be obtained by using of this reaction under standard conditions?

Solution:

According to expression (4.8):

$$A'_{max} = \Delta G$$

At the same time,

$$\Delta G = \Delta H - T\Delta S \qquad\qquad (4.2)$$

Heat of reversible process according to Clausius's equality

$$q_{rev.} = T\Delta S \qquad\qquad (3.21)$$

Let us untroduce expression (3.21) into (4.2):

$A'_{max} = \Delta G = \Delta H - T\Delta S = \Delta H - q_{rev.} = -\ 139.97 - (-\ 14.995) = -\ 124.975$ *kJ/mol.*

Problem № 4.8

The reaction is given:

$$C_2H_2\text{ (g)} + 2\ H_2O\text{ (liq.)} \rightarrow CH_3COOH\text{ (liq.)} + H_2\text{ (g)}$$

Its standard $\Delta H^0 = -139.97\ kJ/mol$. Temperature coefficient of useful work of reversible process under $P = const$ conditions is equal to 50.32 J/K.

1) What are ΔS_{298} and ΔG_{298} of the reaction?

2) What is the difference between A'_{max}, fulfilled at 25^0C and 50^0C?

3) At what temperature proceeds this reaction more easily?

Solution:

1) As is known,

$$-\Delta S = \left(\frac{\partial \Delta G}{\partial T}\right)_P \tag{4.13}$$

but

$$\Delta G = A'_{max} \tag{4.8}$$

Hence

$$\Delta S = -\left(\frac{\partial \Delta G}{\partial T}\right)_P = -\left(\frac{\partial A'_{max}}{\partial T}\right)_P = -50.32\ J\ /\ K \times mol$$

$$\Delta G = \Delta H - T\Delta S = -139970 - 298 \times (-50.32) = -139970 + 14995 =$$

$$= -124975\ J/mol$$

2) Under conditions $P = const$

$$\Delta G_{T_2} - \Delta G_{T_1} = -\Delta S(T_2 - T_1)$$

If take into account, that $\Delta G = A'_{max}$, then

$$A'_{323} - A'_{298} = -\Delta S(323 - 298) = 50.32 \times 25 = 1258 \ J/mol$$

3) As is already seen,

$$\Delta G_{323} - \Delta G_{298} = 1258 \ J/mol$$

Accordingly, the negative value of ΔG decreases with increasing of temperature, which indicates that the considered reaction proceeds with more difficulty at high temperatures.

Problem № 4.9
$\Delta S^0 = 20 \ J / K \times mol$ for reaction of synthesis of hydrogen chloride under standard conditions ($25^0 C$, 1 atm):

$$H_2 + Cl_2 \rightarrow 2 \ HCl$$

Will the useful work (A') of reversible reaction increase or decrease with increasing of temperature to $50^0 C$?

Solution:
Change of ΔG of chemical reaction with temperature is expressed as follows:

$$d\Delta G = -\Delta S dT \qquad (when \ P = const) \qquad (4.13)$$

$$\int_1^2 d\Delta G = -\int_{T_1}^{T_2} \Delta S dT$$

Since range of temperature is narrow, it may be assumed that $\Delta S = const.$ Then

$$\int_1^2 d\Delta G = -\Delta S \int_{T_1}^{T_2} dT$$

$$\Delta G_{T_2} - \Delta G_{T_1} = -\Delta S (T_2 - T_1)$$

$$\Delta G_{50°C} - \Delta G_{25°C} = -20(50 - 25) = -500 \; J/mol$$

As is known, at reversible proceeding of process under $P,T = const$ conditions:

$$\Delta G = A'_{max} \tag{4.8}$$

Thus,

$$\Delta G_{50°C} - \Delta G_{25°C} = A'_{50°C} - A'_{25°C} = -0.5 \; kJ/mol$$

Thus, work fulfilled by synthesis of hydrogen chloride at $50^0 C$ exceeds work fulfilled at $25^0 C$. (Following to our designation, $A'<0$, when system fulfills a work).

Problem № 4.10
$\Delta H^0 = -890.4$ kJ / mol and $\Delta S^0 = -242.6$ J / K × mol for the reaction of combusion of methane:

$$CH_4 + 2\,O_2 \rightarrow CO_2 + 2\,H_2O_{\text{(liq.)}}$$

What is the maximal useful work of reaction at $25^0 C$ and $50^0 C$?

Solution:
In the reversible process, when the maximal useful work is fulfilled:

$$A'_{max} = \Delta G \qquad (P,T = const) \tag{4.8}$$

$$A'_{298} = \Delta G_{298} = \Delta H_{298} - T\Delta S_{298} = -890400 - 298 \times (-242.6) =$$

$$= -890400 + 72295 = -818105 \; J/mol$$

According to expressions (4.8) and (4.13),

$$A'_{323} = \Delta G_{323} = \Delta G_{298} - \int_{298}^{323} \Delta S dT = \Delta G_{298} - \Delta S(323 - 298) =$$

$$= -818105 - (-242.6)(323\text{-}298) = -818105 + 6065 = -812040 \, J/mol$$

Thus, the useful work of methane combustion with increasing of temperature from 25^0C to 50^0C decreases by $\sim 6 \, kJ$.

The contrary effect takes place at synthesis of hydrogen chloride: useful work increases with increasing of temperature (see problem № 4.9). The difference in dependencies of work on temperature for these reactions is stipulated by the opposite signs of its ΔS: $\Delta S < 0$ at combustion of methane, but $\Delta S > 0$ at synthesis of hydrogen chloride.

Problem № 4.11
$\Delta H^0_{298} = -46180$ J/mol and $\Delta G^0_{298} = -16554$ J/mol for the reaction of synthesis of ammonia:

$$0.5 \, N_2 + 1.5 \, H_2 \rightarrow NH_{3 \, (g)}$$

1) Calculate the value of ΔG at 400 and 500^0C. Which of reactions (forward or reverse) proceed spontaneously at 400 and 500^0C?

2) At what temperature is equally probable proceeding of forward and reverse reactions?

Solution:
1) From expression (4.13) follows, that at $P = const$ conditions:

$$d\Delta G = -\Delta S dT$$

Hence

$$\Delta G_{T_2} - \Delta G_{T_1} = -\int_{T_1}^{T_2} \Delta S dT$$

Let us assume, that $\Delta S = const$ in the range $400 \div 500$ K, then

$$\Delta G_{T_2} - \Delta G_{T_1} = -\Delta S(T_2 - T_1)$$

$$\Delta G_{T_2} = \Delta G_{T_1} - \Delta S(T_2 - T_1) \tag{a}$$

ΔS for reaction of synthesis of ammonia is:

$$\Delta S = S_{NH_3} - 0.5 S_{N_2} - 1.5 S_{H_2} = 192.5 - 0.5 \times 192.1 - 1.5 \times 130.6 =$$

$$= -99.45 \; J/K \times mol$$

Let us determine the values of ΔG at 400 and 500 K by using of expression (a):

$\Delta G_{400} = \Delta G_{298} - (-99.45)(400 - 298) = -16554 + 10143 = -6410$
J/mol

$\Delta G_{500} = \Delta G_{298} - (-99.45)(500 - 298) = -16554 + 20089 = +3535$
J/mol

Thus, the reaction of synthesis of ammonia proceeds spontaneously ($\Delta G < 0$) at 298 and 400 K. But this reaction is not spontaneous at 500 K ($\Delta G > 0$). In these conditions the reverse reaction proceeds spontaneously.

2) $\Delta G_{T_x} = 0$ at that temperature (T_x), on which forward and reverse reactions are equally probable. Let us introduce this value into expression (a):

$$\Delta G_{T_x} = \Delta G_{298} - \Delta S(T_x - 298)$$

$$0 = -16554 - (-99,45)(T_x - 298)$$

$$16554 = 99.45 \, (T_x - 298)$$

$$T_x - 298 = 166.45$$

$$T_x = 298 + 166.45 = 464.45 \text{ K}$$

Problem № 4.12

Calculate equilibrium pressure of transformation of graphite into diamond at 298 K and 1400 K by using of following data:

T K	ΔG^0, J/mol	$\Delta \overline{V} = \overline{V}_{diam.} - \overline{V}_{graph.}$, cm^3
298	2903	-1.92
1400	9372	-1.90

Do not take into account change of density with increasing of pressure.

Solution:

Following to expression (4.12):

$$d\Delta G = \Delta V dP \qquad \text{(when } T = const) \qquad (4.12')$$

where $d\Delta G$ is change of ΔG at allotropic transformation, stipulated by change of pressure (dP), and

$$\Delta V = \sum V_{prod.} - \sum V_{react.} \qquad (4.14)$$

According to condition of problem, $d_{diam.} = const$ and $d_{graph.} = const$. It follows, that the molar volumes of diamond and graphite do not change with increasing of pressure and

$$\Delta V = V_{diam.} - V_{graph.} = const$$

Then from expression (4.12') will be obtained:

$$\int_{1}^{2} d\Delta G = \Delta V \int_{1}^{P_{equil}} dP$$

$$\Delta G_2 - \Delta G_1 = \Delta V(P_{equil.} - 1)$$

$$P_{equil} = \frac{\Delta G_2 - \Delta G_1}{\Delta V} + 1 \qquad \text{(a)}$$

where ΔG_1 corresponds to allotropic transformation at 25^0C and 1 atm, but ΔG_2 – at 25^0C and $P_{equil.}$.

According to condition of problem, at 25^0C and $P_{equil.}$ diamond and graphite are presented in equilibrium with each other.. This means, that

$$\Delta G_2 = 0$$

Then it may be obtained from expression (a):

$$P_{equil.} = \frac{0 - \Delta G_1}{\Delta V} = -\frac{\Delta G_1}{\Delta V} \qquad \text{(b)}$$

Let us introduce the values of respective quantities into expression (b):

1) $T = 298$ K; $\Delta G_1 = 2903$ J/mol; $\Delta V = -1.92$ cm^3.

$$P_{equil.} = -\frac{2903 \ J \ / \ mol}{-1.92 \ cm^3 \ / \ mol} = \frac{2903 \times 9.867 \times 10^{-3} \ L \times atm \ / \ mol}{1.92 \times 10^{-3} \ L \ / \ mol} =$$

$$= 14922 \ atm$$

2) $T = 1400$ K; $\Delta G_1 = 9372$ J/mol; $\Delta V = -1.90$ cm^3

$$P_{equil.} = -\frac{9372 \ J \ / \ mol}{-1.90 \ cm^3 \ / \ mol} = \frac{9372 \times 9.867 \times 10^{-3} \ L \times atm \ / \ mol}{1.92 \times 10^{-3} \ L \ / \ mol} =$$

$$= 48671 \ atm.$$

Tabulate the obtained results.

Table 4.2. Dependence of equilibrium pressure of reaction: *graphite* → *diamond* on the temperature

T K	$P_{equilibrium}$, atm
298	14922
1400	48671

As is shown from Table 4.2, most high pressure is necessary for transformation of graphite into diamond.With that, increase of temperature does not facilitate this process (The data in the condition of problem also indicate the last circumstance, viz. increase of ΔG with increasing of temperature).

Appendix A

Formulations of Thermodynamic Laws

Zeroth Law of Thermodynamics*

1. If systems A and B are separately in thermal equilibrium with system C, then systems A and B are in thermal equilibrium with each other.

R. Fowler

2. If any body is in thermal equilibrium with two other bodies simultaneously, then the last two are also in thermal equilibrium with each other.

V. Solyakov

3. If two (or more) systems are in thermal equilibrium with each other, then their temperatures are the same.

F. Daniels, R. Olberty

---

* This law was formulated by R. Fowler in 1931. Zeroth law introduces in the thermodynamics notions of temperature and thermal equilibrium Therefore it must be taken a lead over the other laws. But the principles of thermodynamics were formulated long before that time and change of their numbering was not considered expedient. Because of Fowler's law was called as zeroth law of thermodynamics.

First Law of Thermodynamics

1. The margin of energy of any isolated system is constant.

J. Joule

2. The internal energy of system, presented in the given state, is characterized with one certain value, which is independent of changes undergone by system untill reaching this state.

Y. Gerasymov

3. The internal energy of system represents single–value, continued and finite function of state of the system.

Y. Gerasymov

4. The increment of internal energy represents a total differential of state functon at infinitesimal changes of parameters of system's state.

Y. Gerasymov

5. The change of internal energy of system at transferring from one state to another is independent of the way of process, but depends only on initial and final states of the system.

Y. Gerasymov

6. The internal energy of system represents single–value function of its state and varies as a result of external influence only.

I. Bazarov

7. *Perpetuum mobile* of first kind* is impossible.

W. Ostwald

8. It is impossible to construct such machine, action of which would cause a sole result – production or destruction of mechanical work or energy of any kind.

J. Partington

---

*Perpetuum mobile of first kind – machine, which would work without delivery of energy i.e. would produce a work from nothing.

Second Law of Thermodynamics

1. The heat transfer from the colder body to the hoter one is impossible to be a sole result of any complex of processes.

R. Clausius

2. The heat transfer from the cold system to the hot one is impossible, if any other change will not occur in these systems or surroundings simultaneously.

P. Epstein

3. If heat transfers from body *A* to body *B* by contact, it is impossible the existence of such process, a sole result of which will be a heat transfer from *B* to *A*.

E. Fermi

4. The spontaneous transfer of energy in the form of heat from the cold body to the hoter one is impossible.

V. Solyakov

5. The process of thermal conductivity is irreversible.

Y. Gerasimov

6. It is impossible to transfer heat of the body into work, if any other influence except the cooling will not be performed on it.

V. Kelvin

7. Such circular process does not exist, a sole result of which will be the gaining of heat from heater and fulfilment of work.

Kelvin—Planck

8. The spontaneous process of transformation of work into heat via the friction is irreversible.

Y. Gerasymov

9. It is impossible to remove heat from a body and transform this heat into the work, if any other change does not occur in the system or surroundings.

P. Epstein

10. The mechanical work may be completely transformed into heat, but the transformation of heat into work will be undoubtedly incomplete, because always, when a certain amount of heat is transformed into the work, the other amount of heat must essentially undergo the respective compensating change.

M. Planck

11. It is impossible to construct such machine, which would work circulary and a sole result of its action would be the gaining of heat from reservoir and the lifting of load.

J. Partington

12. The heat of the coldest body among others, taking part in circular process, can not serve as a source of work.

V. Kelvin

13. *Perpetuum mobile* of second kind* is impossible.

W. Ostwald

14. It is impossible to construct such machine, which would work at the expense of some source (e.g. atmosphere, ocean) and would have temperature the same as the source itself.

V. Kelvin

15. It is impossible to construct thermal machine with only one source of the heat.

I. Prigogine

---

*Perpetuum mobile of second kind – machine, which completely transforms heat into work without refrigerator.

16. Not a machine fulfil a work without transferring of heat to the low–temperature source.

J. Fenn

17. Not a thermal machine may have efficiency which exceeds efficiency of Carnot's cycle

J. Fenn

18. The negative (nonspontaneous) process is impossible to be a sole result of cycle.

Clausius–Kelvin

19. Entropy of adiabatic system is invariable in the equilibrium processes and increases in the nonequilibrium one.

Y. Gerasymov

20. Entropy of isolated system does not change in the equilibrium processes and increases in the nonequilibrium one.

Y. Gerasymov

21. The total entropy of all considered bodies in any reversible process increases.

G. Lewis

22. If the system contains all of that bodies which undergo the transformation, the entropy of system increases in the natural processes.

M. Saha

23. Any spontaneous process in any isolated system results in the increasing of entropy.

J. Fenn

24. Entropy of the universe always increases.

J. Fenn

25. Energy of the universe is constant, entropy of the universe tends to maximum.

R. Clausius

26. Entropy is induced frequently, but never is destructed.

J. Fenn

27. Any system set at liberty, changes in the direction of maximal probability.

G. Lewis

28. Only these processes may proceed spontaneously, in which the system transfers from less probable state into the more probable one.

V. Solyakov

29. Some phenomena proceed in the certain direction not by reason that their proceeding in the other direction is impossible, but because of minute probability of their proceeding in the latter direction.

M. Gardner

30. It is impossible to reduce entropy of system of bodies without changes in other bodies, presented in contact with system.

M. Planck

31. The law of monotonous increasing of entropy – the second principle of thermodynamics – holds a prominent place between laws of the nature…If it is revealed, that your theory is in contrast with second principle of thermodynamics, no hopes are remained with you: your theory is doomed to ignominious end.

A. Eddington

32. The equilibrium in isolated system is globally stable and is characterized with maximum of entropy.

I. Prigogine

33. An universal tendency to the degradation of mechanical energy exists in the nature.

V. Kelvin

34. The energy tries for dissipation in the closed system i.e. it attempts to transfer into evenly distributed thermal energy; in other words, it undergoes a degradation or depreciation.

V. Kelvin

35. Energy tends to the dissipation.

P. Atkins

36. Energy of the universe is constant, but the possibility of its use reduces with the increasing of entropy of universe. Entropy in relation to energy plays the same part, as inflation towards currency i.e. devaluates it. Thus, if energy represents an ability to perfom work, entropy is a measure of devaluation of this ability.

J. Fenn

37. Only processes of dissipation and degradation exist primordially. The universe is overflowed by waves of chaos, which have no reason and explanation. This process does not possess the primordial purpose, only the continuous motion exists in it.

P. Atkins

38. Energy is dissipated everywhere and always. But certain structures are born in this process of degradation. Although they are transient, but some among them can exist during million years.

P. Atkins

39. According to classical formulation of the second law of thermodynamics, the thermal death (maximum of entropy) expects us in the future. But it becomes obvious at present: such a great amount of entropy was arisen at a origin of the universe, that danger of thermal death no more exists.

I. Prigogine

Third law of Thermodynamics

1. The entropy of individual ideal crystalline substance is zero at zero Kelvin.

M. Planck

2. Absolute zero of temperature is inaccesible.

M. Planck

Appendix B

Units

Basic SI Units

Physical Quantity	Name of Unit	Symbol
Length	meter	m
Mass	kilogram	kg
Time	second	s
Electric Current	ampere	A
Thermodynamic Temperature	kelvin	K
Quantity of Substance	mole	mol
Luminosity	candela	cd

Derivative SI Units

Physical Quantity	Name of Unit	Dimension
Force	newton (N)	$kg \cdot m /s^2$
Pressure	pascal (Pa)	N / m^2
Energy	joule (J)	$N \cdot m$
Power	watt (wt)	J / s

Ratio between Units

$1L = 10^3 \text{ cm}^3 = 1\text{dm}^3 = 10^{-3} \text{ m}^3$
$1\text{atm} = 1 \text{ bar} = 760 \text{ torr} = 1.01325 \cdot 10^5 \text{ Pa} = 1.013 \cdot 10^6 \text{ dyne / cm}^2$
$1N = 10^5 \text{ dyne} = 10^5 \text{ g} \cdot \text{cm / s}^2$
$1\text{wt} = 1 \text{ J / s}$
$1\text{horsepower (hp)} = 735 \text{ wt} = 0.735 \text{ kwt}$

$T \text{ K} = 273.15 + t^0 \text{C}$	$t^0 \text{C} = T \text{ K} - 273.15$
$T \text{ K} = 255.37 + 5/9 t^0 \text{F}$	$t^0 \text{C} = 5/9 \; t^0 \text{F} - 17.78$
$T^0 \text{F} = (t^0 \text{C} + 17.78) \cdot 9/5$	$t^0 \text{F} = (T \text{ K} - 255.37) \cdot 9/5$

1 inch = 2.54 cm
1 foot = 12 inches = 30.479 cm
1 gallon = 3.785 L
1 pound = 453.6 g
Volume of 1 mole ideal gas at s.t.p. (0^0 C and 1atm) $= 22.415 \cdot 10^{-3} \text{ m}^3 /$ mol $= 22.415$ L /mol.

Ratio between Energy Units

UNIT	joule	erg	calorie	liter x atmosph	kilogram x meter	kilowatt x hour
joule	1	10^7	0.239	$9.867 \times \times 10^{-3}$	0.102	$2.778 \times \times 10^{-7}$
erg	10^{-7}	1	$2.39 \times \times 10^{-8}$	$9.867 \times \times 10^{-10}$	$1.02 \times \times 10^{-8}$	$2.778 \times \times 10^{-14}$
calorie	4.184	$4.184 \times \times 10^7$	1	$4.129 \times \times 10^{-2}$	0.427	$1.162 \times \times 10^{-6}$
liter x atmosph	$1.0134 \times \times 10^2$	$1.0134 \times \times 10^9$	24.22	1	10.337	$2.815 \times \times 10^{-5}$
kilogram x meter	9.8067	$9.8067 \times \times 10^7$	2.342	$9.676 \times \times 10^{-2}$	1	$2.724 \times \times 10^{-6}$
kilowatt x hour	3.6×10^6	3.6×10^{13}	$8.604 \times \times 10^5$	3.552×10^4	$3.672 \times \times 10^5$	1
R	8.3143	$8.3143 \times \times 10^7$	1.98725	$8.2057 \times \times 10^{-2}$	0.8478	$2.31 \times \times 10^{-6}$

Fundamental Constants

Atomic mass unit	u	$1.66057 \cdot 10^{-27}$ kg
Avogadro's number	N_A	$6.02 \cdot 10^{23}$ mol^{-1}
Planck's constant	h	$6.62618 \cdot 10^{-34}$ J $\cdot$ s
Boltzmann's constant	k	$1.38066 \cdot 10^{-23}$ J / K

Numerical Values of Some Quantities

$\pi = 3.14159$

$e = 2.71828$

$lge = 0.434$

$ln10 = 2.3$

$lnx = 2.3lgx$

Prefixes

pico	*nano*	*micro*	*milli*	*centi*	*deci*
10^{-12}	10^{-9}	10^{-6}	10^{-3}	10^{-2}	10^{-1}

deka	*hecto*	*kilo*	*mega*	*giga*	*tera*
10	10^2	10^3	10^6	10^9	10^{12}

Designation

The molar value of any quantities is symbolized so: $\overline{C}, \overline{S}, \overline{\Delta U}, \overline{\Delta H}, etc.$

Index